D1327109

Genetic Mapping of Disease Genes

Genetic Mapping of Disease Genes

Edited by

Ivar-Harry Pawlowitzki
Institute of Human Genetics, University of Muenster,
Vesaliusweg 12–14, 48149 Munster, Germany

J.H. Edwards
Department of Biochemistry,
University of Oxford, South Parks Road,
Oxford OX1 3QU, UK

Elizabeth A. Thompson
Department of Statistics, University of Washington,
Box 354322, Seattle, WA 98195, USA

ACADEMIC PRESS
Harcourt Brace and Company, Publishers
San Diego   London   Boston   New York
Sydney   Tokyo   Toronto

Copyright © 1997 by ACADEMIC PRESS
Except the chapter titled The early history of the statistical estimation of linkage,
copyright © 1996 Annals of Human Genetics

Academic Press, Inc.
525 B Street, Suite 1900, San Diego, California 92101–4495, USA
http://www.apnet.com

Academic Press Limited
24–28 Oval Road, London, NW1 7DX, UK
http://www.hbuk.co.uk/ap/

ISBN 0-12-232735-7

Genetic mapping of disease genes / edited by Ivar-Harry Pawlowitzki,
John H. Edwards, Elizabeth A. Thompson.
    p. cm.
    Includes index.
    ISBN 0-12-232735-7 (alk. paper)
    1. Genetic disorders.   2. Gene mapping.   I. Pawlowitzki, Ivar-Harry.
II. Edwards, J.H. (John Hilton)   III. Thompson, E.A.
(Elizabeth Alison), 1949–   .
    [DNLM: 1. Hereditary Diseases––genetics.   2. Chromosome Mapping.
QZ 50 G3255 1997]
RB155.5.G458   1997
616′.042––dc21
DNLM/DLC
for Library of Congress   97–11844
                                                                     CIP

A catalogue record for this book is available from the British Library

Typeset by Phoenix Photosetting, Chatham, Kent
Printed in Great Britain by
WBC Book Manufacturers, Bridgend, Mid Glamorgan

97 98 99 00 01 02 EB 9 8 7 6 5 4 3 2 1

To
Angelika
(b. 17.2.1944, d. 17.6.1996)
From I.H.P.

Preface

This book is based on contributions to a meeting sponsored by the European Union and held in Oxford, UK. The venue was chosen to honour the life work of John Edwards, one of the co-editors of this book on the occasion of his retirement. John Edwards' name has been immortalized in medical terminology due to his early cytogenetic studies; in addition he has contributed in a major way to the methodology of linkage analysis of genes over the last several decades.

The Oxford meeting on *Genetic Mapping of Disease Genes* turned out to be extraordinary for a variety of reasons. First, numerous researchers who have made a fundamental impression on the methodology of genetic mapping accepted our invitation. A gathering of so many leading scientists in this research field was surely a unique occasion. In addition, the shortcomings of established mapping procedures in diseases other than monogenic were discussed and new methodological approaches to the mapping of genes underlying oligogenic and polygenic diseases as well as in autosomal recessive disorders with considerable non-allelic heterogeneity were presented. Furthermore, much of the material presented was hitherto unpublished. Finally, a broad overview on the state of the art in genetic mapping of disease genes resulted from the sum of papers. As Elizabeth Thompson said in her concluding remarks 'The integration of theory, algorithms, tools, and applications achieved by this meeting will advance both theory and practice.'

For these reasons it seemed desirable, if not imperative, to publish this book, which contains the updated versions of most of the contributions to the Oxford meeting.

Münster, Germany, November 1996   *I.H. Pawlowitzki*

Acknowledgements

The meeting on 'Genetic Mapping of Disease Genes' held in Oxford, UK, St Catherine's College, December 15–17, 1995 and part of the editorial work were supported by the European Union (Directorate General XII, Grant Nos. BMH1-CT92-1789 and BMH4-CT96-0032 to I.H.P.). I. Craig, J.H. Edwards (Oxford), and B. Müller–Myhsok (Hamburg) were co-organizers of the meeting. Markus Preising, F. Wortmann, and E. Hobbeling efficiently assisted in organizing the meeting in Oxford and in technical work related to the editing of this book.

Contents

.

SECTION V: APPLICATIONS

SECTION VI: FINAL COMMENTS

Contributors

Baur, Max P.
Inst. f. Med. Statistik, Universitätskliniken
Sigmund-Freud-Str. 25, 53105 Bonn, FRG

Bhattacharya, Shomi. S.
Department of Molecular Genetics, Institute of Ophthalmology, University
College London, 11–43 Bath Street, London EC1V 9EL, UK

Bruford, Elspeth A.
MRC Human Genetics Unit, Western General Hospital
Crewe Road, Edinburgh EH4 2XU, UK

Carothers, Andrew D.
MRC Human Genetics Unit, Western General Hospital
Crewe Road, Edinburgh EH4 2XU, UK

Clerget-Darpoux, Françoise
Institut National de la Recherche Médicale (U155), Château de Longchamp
Bois de Boulogne, F-75016 Paris, France

Davis, Sean
University of Pittsburgh, Department of Human Genetics
Crabtree Hall, Room A310, Pittsburgh, PA 15261, USA

Devoto, Marcella
Laboratory of Statistical Genetics, The Rockefeller University
1230 York Avenue, Box 192, NY 10021-6399, USA

Dicks, Jo[1]
Genetics Laboratory, Department of Biochemistry
South Parks Road, Oxford OX1 3QU, UK

Edwards, A.W.F.
Gonville and Caius College,
Cambridge CB2 1TA, UK

---

[1] Present address: John Innes Centre, Norwich Research Park, Colney, Norwich NR4 7UH, UK.

Edwards, J.H.
Department of Biochemistry, University of Oxford
South Parks Road, Oxford OX1 3QU, UK

Evans, K.
Department of Molecular Genetics, Institute of Ophthalmology
11–43 Bath Street, London EC1V 9EL, UK

Goldin, Lynn R.
Clinical Neurogenetics Branch, National Institute of Mental Health
Bethesda, MD 20892, USA

Grimm, Tiemo
Abteilung für Medizinische Genetik im Institut für Humangenetik,
Universität Würzburg / Biozentrum
Am Hubland, 97074 Würzburg, FRG

Knapp, Michael
Inst. f. Med. Statistik, Universitätskliniken
Sigmund-Freud-Str. 25, 53105 Bonn, FRG

Leal, Suzanne M.
Laboratory of Statistical Genetics, The Rockefeller University
1230 York Avenue, New York, NY 10021-6399, USA

Lio, Pietro
Human Genetics Department, Level G, Princess Anne Hospital
Coxford Road, Southampton SO16 5YA, UK

Morton, Newton E.
Human Genetics Department, Level G, Princess Anne Hospital
Coxford Road, Southampton SO16 5YA, UK

Müller-Myhsok, Bertram
Bernhard-Nocht-Institut für Tropenmedizin, Abt. f. Molekulargenetik
Bernhard-Nocht-Str. 74, 20359 Hamburg, FRG

Ott, Jurg
Laboratory of Statistical Genetics, The Rockefeller University
1230 York Avenue, New York, NY 10021-6399, USA

Smith, Cedric A.B.
The Galton Laboratory, Department of Genetics and Biometry
4 Stephenson Way, London NW1 2HE, UK

Stephens, David A.
Imperial College, Department of Mathematics
Huxley Building, 180 Queen's Gate, London SW7 2BZ, UK

te Meerman, Gerard J.
University of Groningen, Department of Medical Genetics
Antonius Deusinglaan 4, NL-9713 AW Groningen, The Netherlands

Teague, Peter W.
MRC Human Genetics Unit, Western General Hospital
Crewe Road, Edinburgh EH4 2XU, UK

Thompson, Elizabeth A.
Department of Statistics, University of Washington
Box 354322, Seattle, WA 98195, USA

van der Meulen, Martin A.[1]
University of Groningen, Department of Medical Genetics
Antonius Deusinglaan 4, NL-9713 AW Groningen, The Netherlands

Visser, Frank
HGMP Resource Centre, Hinxton Hall
Hinxton, Cambridge CB10 1RQ, UK

Weeks, Daniel Eastman
University of Pittsburgh, Department of Human Genetics
Crabtree Hall, Room A310, Pittsburgh, PA 15261, USA
and
The Wellcome Trust Centre for Human Genetics, University of Oxford
Windmill Road, Oxford OX3 7BN, UK

Wright, Alan F.
MRC Human Genetics Unit, Western General Hospital
Crewe Road, Edinburgh EH4 2XU, UK

---

[1] Present address: University of Groningen, Department of Psychiatry, PO Box 30-001, NL-9700 RB, Groningen, The Netherlands.

SECTION I

INTRODUCTION AND HISTORY

Chapter 1
Introduction
J.H. Edwards

First may I say how pleased I am that Oxford has been chosen for this venue. It can claim a major contribution to the study of linkage in the past and we may anticipate a continuing contribution in the future, especially in the fields of genetic predispositions to infectious and autoimmune disease, including malaria and diabetes mellitus, and, through the nearby Medical Research Council Radiobiology Unit at Harwell, in the detailed decipherment of linkage in the mouse and homologies with human disease.

This year, 1995, is a good year for anniversaries, with Haldane, Sprunt and Haldane's work on mice in Oxford in 1915 (Haldane *et al.*, 1915), the development of practical methods of estimation when counting was impractical by Morton in Wisconsin in 1955 (Morton, 1955), the development of *in vitro* hybridization by Harris and Watkins (1965) and the direct visualisation of nuclear variants rather than their products by the 'Southern blot' in Edinburgh 1975 (Southern 1975).

In 1965 Harris and Watkins demonstrated the feasibility of culturing cells hybridised from different species. Since these cells 'lost' various chromosomes from one of the parent species, and most enzymes from different species differ sufficiently to be distinguished on electrophoresis, this allowed genes to be localised to chromosomes, and, if translocations were present, to parts of chromosomes. The genetic mapping, which exploits the normal breakage and joining of chromosome at the production of germ cells, made feasible by Morton's work, could now be supplemented by 'physical mapping' to produce two related but formally incommensurate maps.

In 1915 Haldane, a fellow of New College, published a paper usually, but

GENETIC MAPPING OF DISEASE GENES
ISBN 0-12-232735-7

erroneously, considered the first paper on linkage in mammals. Last January we celebrated its eightieth anniversary in the presence of one of its coauthors, Haldane's sister, Lady Naomi Mitchison. At least this paper can claim to be the first case of disturbed segregation in disorders manifest in the mammalian eye, albinism and pink-eye dilute, even if interpreted as 'reduplication.' An appropriate disorder, as Spooner, the Warden of New College where Haldane was a fellow, was an albino. Spooner's name is preserved by the reputed habit, which he always denied, of transposing consonants.

Later, after the interruption of war service in France, Mesopotamia and India, Haldane published a second major paper (Haldane, 1919). The first, written when on leave from France, started 'one of us is already dead', and it is unlikely he was adequately informed of the major work from the 'fly room' in New York during the next three years. In a second paper in 1919 he distinguished structural linkage related to physical position from functional linkage, which reflected it but was only manifest through recombinant events that could occur more than once and cancel out. He defined the unit of functional length as a 'Morgan' indirectly through the centimorgan, a distance conferring a 1% chance of recombination.

This followed the key papers of Sturtevant (Sturtevant, 1913) on how to order loci and Robbins in 1918 (Robbins, 1918) on linkage inequality, or allelic association secondary to the effects of close linkage in distant relatives who form populations, a phenomenon often termed linkage disequilibrium, in spite of the historically correct term linkage inequality and the confusion of Fisher's earlier use of 'linkage equilibrium' for the same phenomenon, which he assumed was stabilised by selection. After Fisher's paper of 1922 (Fisher, 1922) on how to position loci once order was defined the major problems of linkage analysis in experimental crosses and some large and complete human families were resolved. The efficient analysis of smaller and fragmented families was only resolved with the help of computers to tabulate lod scores by Morton in 1955 and also by procedures developed and extended by Renwick to 'tailor made' programs for individual families and by myself to family units of parents with grandparents and children, some missing. In 1974 these were generalised and extended by Ott (Ott, 1974) to families of arbitrary size. These developments had little impact on mapping at that time, and most of the major advances made in the next 20 years were through Morton's tables. Renwick (1969) provided a useful summary. These procedures all follow the approach advanced by Haldane and Smith in 1947 (Haldane and Smith, 1947). Ott's approach, which included the 'peeling' algorithm of Elston and Stewart (Elston and Stewart, 1971), was extended by Lathrop and others to cover several loci. Provided that only one gene was related to a disorder, the marker loci used were not too close, so that there was no substantial allelic association and the loci had few alleles whose frequencies were known, the locus with the offending gene could be positioned efficiently by well-tested programs. These

4

methods have provided most of the linkages established in the last twenty years. However, the advance in the numbers of both marker loci, and their alleles, as well as the interacting assumption that all mutations are different, now reduce the adequacy of the approximations involved.

When Professor Pawlowitzki asked me if I would assist in the local organisation of this conference I agreed, but pointed out that the diagnostic difficulties in ophthalmology meant that common manifestations of genetic disease of the eye could not be classified with sufficient precision beyond stating that the front was going white or the back going black, and it was clear that many loci were involved, some dominant, some recessive, and some X-linked. This meant it was often not possible to use the standard programs that had been so powerful in disorders more amenable to precise diagnosis. The use of these powerful methods in eye disease continues to be applied, but we now need to consider their limitations, to develop alternative approaches, and to relate the strategy for collecting data to the efficiency of various subsets of families and the optimal populations for their study. The simple approach, useful in the past, of collecting the largest families possible and centralising analysis, is highly inefficient in these disorders. It can lead to both loss of power in detection and erroneous confidence limits, the major locus being dragged outside the prescribed range by its rarer brethren. Recessive disorders, and disorders related to several alleles of low penetrance, such as glaucoma, need small families: even the smallest families, that is single individuals and unrelated controls, can be powerful and unbiased, and sufficient given strong candidate loci.

Linkage analysis offers a major approach to diseases of the eye, which like the ear, the brain and the pancreas cannot be biopsied with impunity and must be studied by such indirect approaches as analysing blood and relating the results to genome maps and to maps of related species in which experimental crosses are possible with scheduled autopsies, both before and after birth. Even if the phenotype during life is difficult to define at a precision sufficient to distinguish the related loci, the extreme specialisation of the lens and retina provides a range of strong candidate loci denied to any other organ.

In the past the resources at our disposal were limited to a few markers and a few large families, which could reasonably be assumed to have a single locus: for these the methods of analysis available were efficient and the 'rule of three', that is the use of a lod of three as an arbitrary threshold of credulity, was a useful compromise between missing what was there and finding what was not, while the recommended 10:1 confidence range could be expected to embrace reality. However, the resolving power was limited to at least five megabases, and concentrating the segment of interest from the 3000 megabases present even a thousandfold was inadequate to provide a segment short enough to sequence and explore for the offending segment. Its main use was to separate disorders of similar phenotype, as done by Morton in 1956 (Morton, 1956), and suggest the homology of similar disorders in other species.

Matters have now changed. The numerous loci with multiple alleles at our disposal have imposed problems on methods that assume families can be analysed together on the assumptions that there is only one locus, that the allele proportions are adequately estimated and constant over the populations analysed, that the loci are sufficiently distant for their alleles not to be associated, and that all mutations are different. We need to learn from both the successes of the past and the limitations of the present. Computers have not only advanced in their capacity to do the repetitive arithmetic underlying linkage programs, but also in software enabling them to read and write text in readily intelligible formats, to be interactive, and, in particular, to present detailed and coloured displays. In the past the emphasis on linkage analysis was to inform investigators on how surprised they should be by comparing observations with expectations on some null hypothesis. It is now possible to display suggestive associations and linkages directly on the retina, allowing their interpretation by those with relevant experience aided by such other evidence as is available. This should complement, rather than being merely subservient to statistical guidelines based on systematic exploitation of the ignorance implicit in a null hypothesis. We can use the retina to study the retina. We have the example of the efficient exploitation of such procedures in other species, especially *Drosophila melanogaster* and *Caenorhabditis elegans*, the fly and the worm.

The fortieth anniversary of Morton's development of lod scores is an appropriate time for reassessing the contribution to be expected from linkage and the optimal selection of populations, of families, of laboratory techniques, and of analytical methods. They can no longer be considered in isolation. Nor can they wait for the bicentenary: too much blood and gold is being applied without discussion on strategy or tactics under the erroneous belief that there is no problem and that the procedures of the past merely need extending or applying to faster computers to meet the challenge of the future.

References

Elston, R.C. & Stewart, J. A general model for the genetic analysis of pedigree data. (1971) *Hum. Hered.* **21 (6):** 523–542.

Fisher, R.A. The systematic location of genes by means of crossover observations. (1922) *Am. Naturalist* **56:** 406–411.

Haldane, J.B.S. The combination of linkage values, and the calculation of distances between the loci of linked factors. (1919) *J. Genet.* **8:** 299–309.

Haldane, J.B.S. & Smith, C.A.B. A new estimate of the linkage between the genes for colour-blindness and haemophilia in man. (1947) *Ann. Eugen.* **14:** 10–31.

Haldane, J.B.S., Sprunt, A.D., Haldane, N.M. Reduplication in mice. (1915) *J. Genet.* **5:** 133.

Harris, H. & Watkins, J.F. Hybrid cells derived from mouse and man: artificial hetero-karyous of mammalian cells from different species. (1965) *Nature* **205:** 640–646.

Morton, N.E. Sequential tests for the detection of linkage. (1955) *Am. J. Hum. Genet.* **7:** 277–318.

Morton, N.E. The detection and estimation of linkage between the genes for elliptocytosis and the Rh blood type. (1956) *Am. J. Hum. Genet.* **8:** 80–96.

Ott, J. Estimation of the recombination fraction in human pedigrees: efficient computation of the likelihood for human linkage studies. (1974) *Am. J. Hum. Genet.* **26 (5):** 588–597.

Renwick, J.H. Progress in mapping human chromosomes. (1969) *Brit. Med. Bull.* **25(1):** 65–73.

Robbins, R.B. Some applications of mathematics to breeding problems. III. (1918) *Genetics* **8:** 299.

Southern, E.M. Detection of specific sequences among DNA fragments separated by gel electrophoresis. (1975) *J. Mol. Biol.* **98 (3):** 503–517.

Sturtevant, A.H. The linear arrangement of six sex-linked factors in *Drosophila*. (1913) *J. Exp. Zool.* **14:** 45–59.

Chapter 2
The early history of the statistical estimation of linkage

A.W.F. Edwards

The first observations of non-independent segregation between Mendelian loci were made in the sweet pea by Bateson, Saunders and Punnett (1905, 1906). The hypothesis that such associations were due to the loci being close together on the same chromosome was advanced by Morgan (1911). Generally speaking the early data consisted of 'F_2' segregations from the mating of 'F_1' multiple heterozygotes $AB/ab \times AB/ab$, where A and B were dominant to a and b respectively. The 'gametic ratios' could therefore not be estimated directly, but had to be estimated from the observed 'zygotic ratios'. However, Morgan and his colleagues, working with *Drosophila*, were able to use backcrosses $AB/ab \times ab/ab$ and thus observe gametic ratios directly.

Before 1912 no statistical theory had been developed to estimate the amount of 'coupling' or 'repulsion' (i.e. linkage) from the numbers in the four classes of zygote, but in that year three papers set the scene for future developments. Collins (1912) proposed that Yule's (1900) measure of association

$$Q = (AB.ab - Ab.aB)/(AB.ab + Ab.aB), \qquad \text{(Eq. 2.1)}$$

whose standard error was known, should be calculated from the zygotic data and compared with the values which corresponded to particular gametic ratios. In a second paper, Harris (1912) proposed the use of Pearson's (1900) χ^2 goodness-of-fit to test the observed zygotic ratios against various hypothetical ratios, while a third paper whose idea was to have a profound influence on linkage analysis was

GENETIC MAPPING OF DISEASE GENES   Copyright © 1996 *Annals of Human Genetics* **60**: 237–249
ISBN 0-12-232735-7   All rights of reproduction in any form reserved

written by the undergraduate R.A. Fisher introducing what came later to be called the method of maximum likelihood (Fisher, 1912).

In 1914 the first statistical estimate proper was made by Yule and Engledow (1914), carrying on where Collins (1912) had left off, and published as Engledow and Yule (1914, 1915). In order to estimate the 'coupling-ratio' they worked with a parameter p in terms of which the gametic ratio was $p:\frac{1}{2} - p:\frac{1}{2} - p:p$
and then estimated p, for which purpose they invented the statistical estimation procedure known as the method of minimum χ^2 and applied it to the segregation in the F_2 generation. Now p is simply half the recombination fraction, though Engledow and Yule omitted to mention the fact. They showed that the minimisation of χ^2 led to an equation of the fourth degree, and as an initial approximation they suggested

$$p^2 = 0.25 \, (AB + ab - Ab - aB)$$

which they obtained as the simple least-squares solution. This 'additive method' was later attributed to Emerson (1916). The third estimation procedure to be proposed was put forward by Bridges (1914) and amounted to equating the observed value of Q to its value on replacing the observed frequencies by their expectations as functions of the recombination fraction, and solving for the recombination fraction. It was soon realised that this reduced to a simple 'product method' in which the observed value of the 'cross-product ratio' $AB.ab/Ab.aB$ is set equal to its value when expectations are inserted.

In 1916 Muller (1916a,b,c,d) wrote a four-part report and review *The mechanism of crossing-over* based on his doctoral thesis which included some mathematical considerations. If y_1, y_2 and y_{1+2} are the recombination fractions in segments AB, BC and AC respectively, where A, B and C are three loci in that order on the chromosome, and if crossing-over is a random process, then, he noted:

$$y_{1+2} = y_1(1 - y_2) + y_2(1 - y_1)$$

and thus:

$$y_{1+2} = y_1 + y_2 - 2y_1y_2.$$

Muller went on to consider non-independence between recombinant events: 'In a sense, then, the occurrence of one crossing-over interferes with the coincident occurrence of another crossing-over in the same pair of chromosomes, and I have accordingly termed this phenomenon "interference".' He defined the 'coincidence' as the ratio of the observed to the expected number of cross-overs on the random hypothesis, as suggested by Bridges (1915).

Muller's review showed that in Morgan's group at Columbia University there was a clear understanding that a linkage map constructed by adding the recombination values for adjacent short intervals did not give the recombination values for the longer intervals even in the absence of interference, and also that they possessed the relation which a recombination fraction bore on the random hypothesis to the recombination fractions of any two segments into which the whole was divided. Perhaps they even possessed 'Haldane's mapping function' (Haldane 1919b) relating the map distance to the recombination fraction, for Muller (1916a) wrote 'In the case of higher percents of separation (long distances), the highest of the three frequencies (let us call it *AC*) falls short of the sum of the other two (*AB+BC*), and so it is a smaller number than the distance representing it on the diagram, but it nevertheless (within the normal limits of error) can be calculated from this diagram distance *AC*, for a constant relation was discoverable between this hypothetical distance and the actual frequency.'

Haldane (1919a), in the first of two papers often wrongly supposed to be the foundation of linkage estimation theory, set out 'to obtain from an observed zygotic series the most accurate possible estimate of the corresponding gametic series'. His solution is an inelegant version of Engledow and Yule's (1914) minimum-χ^2 solution of five years before; inelegant because instead of deriving the quartic exactly and using standard methods for its solution, as they did, he makes an approximation which leads him to an iterative solution, which he says gives the same answer as Bridges's (1914) product method anyway. His suggested initial approximations are precisely those put forward by Emerson (1916). Haldane's second paper (1919b) goes over some of the same ground as Muller's review, to which he does not refer, and is famous for the introduction of the mapping function, which, as mentioned above, Muller might already have possessed. A search has so far failed to reveal a copy of Muller's thesis itself, perusal of which might settle the question.

Fisher's youthful paper of 1912 bore fruit in 1921 when he defined likelihood, and then, in 1922a, elaborated the theoretical advantages of his newly-christened method of maximum likelihood. Fisher (1922a) then saw that linkage estimation presented a wonderful opportunity to apply the method. Tackling eight sex-linked *Drosophila* loci whose order he assumed known he wrote down the likelihood function for the seven recombination fractions, followed by the corresponding equations for the maximum, which he found by differentiating the log-likelihood. These equations were then linearised and solved to give approximate maximum-likelihood estimates. He explicitly ignored the possibility of multiple crossing-over so that he was able to assume additivity for the recombination fractions. J.H. Edwards (1989) has reworked Fisher's example and derived the exact maximum-likelihood solution.

1925 saw the publication of the first edition of Fisher's famous book *'Statistical Methods for Research Workers.'* In the introductory chapter Fisher discussed the need for efficient estimation procedures, and gave as an example of the method of maximum likelihood the estimation of the recombination fraction from F_2 data. For the second edition (1928) this section was expanded into a new chapter *'The Principles of Statistical Estimation.'* The same material, treated more expansively, appeared in a joint paper in the same year (Fisher and Balmukand, 1928).

Fisher and Balmukand considered five ways of estimating the recombination fraction:

(1) the additive method (Engledow and Yule, 1914; Emerson, 1916);
(2) a modified additive method, with different weighting, invented by Fisher for the purpose of illustration;
(3) the product method (Bridges, 1914);
(4) the method of maximum likelihood (Fisher, 1922b);
(5) the method of minimum χ^2 (Engledow and Yule, 1914).

They found the estimated recombination fraction by each method, and then showed how to compute the sampling variance of each. Knowing that the method of maximum likelihood (4) and the method of minimum χ^2 (5) gave the smallest possible sampling variance (which Haldane, 1919a, had correctly obtained), and were therefore fully efficient, they demonstrated that the product method (3) also belonged to this class of efficient methods but that the additive methods (1) and (2) gave variances which showed that they were not fully efficient except at special values of the recombination fraction.

By 1928 Fisher had thus exhibited through the example of linkage what his theoretical researches had revealed; namely the superiority of the method of maximum likelihood. Henceforth the use of any method but maximum likelihood would have to be justified on different grounds, such as the robustness of the product method in the presence of differential viability, or ease of computation. Fisher's theoretical advances had enabled him to assess all the developments of the previous 14 years, and to place the theory of linkage estimation on a firm footing.

It was a stroke of good fortune or perhaps an indication of the rapidity of scientific progress at the time, that the decade which saw the chromosomal theory of the gene firmly established by Morgan was followed immediately by the decade which saw the mathematical foundations of theoretical statistics firmly established by Fisher. The analysis of the *Drosophila* linkage data (Fisher, 1922b) was the very first application of the method of maximum likelihood. Not only were the geneticists fortunate in being handed the most advanced statistical techniques, but the statisticians were fortunate in being able to hone these techniques on such ideally-suited data. By 1928 the statistical (but not the computational) problems of linkage estimation in experimental organisms had thus been solved, leaving Haldane (1934) and Fisher (1934, 1935) to turn to the peculiar problems of linkage estimation in man, where a start had already been made by other workers in Germany and England.

In the history of those two decades it should, however, be noted that a fundamental contribution by Engledow and Yule (1914, 1915) was inexplicably overlooked, and that Morgan's group, largely through the work of Muller, were by 1919 certainly in possession of a mapping function, presumably the one associated with Haldane's name.

Acknowledgement

Condensed from the full version of the paper in *Annals of Human Genetics* Volume **60**: 237–249 (1996) with the permission of the Editors.

References

Bateson, W., Saunders, E.R., Punnett, R.C. (1905) *Reports to the Evolution Committee of the Royal Society, II. Experimental Studies in the Physiology of Heredity*. The Royal Society, London.

Bateson, W., Saunders, E.R., Punnett, R.C. (1906) *Reports to the Evolution Committee of the Royal Society, III. Experimental Studies in the Physiology of Heredity*. The Royal Society, London.

Bridges, C.B. The chromosome hypothesis of linkage applied to cases in sweet peas and primula. (1914) *Am. Naturalist* **48**: 524–534.

Bridges, C.B. A linkage variation in *Drosophila*. (1915) *J. Exp. Zool.* **19**: 1–21.

Collins, G.N. Gametic coupling as a cause of correlations. (1912) *Am. Naturalist* **46**: 569–590.

Edwards, J.H. The locus positioning problem. (1989) *Ann. Hum. Genet.* **53**: 271–275.

Emerson, R.A. The calculation of linkage intensities. (1916) *Am. Naturalist* **50**: 411–420.

Engledow, F.L., Yule, G.U. The determination of the best value of the coupling-ratio from a given set of data. (1914) *Proc. Camb. Phil. Soc.* **17**: 436–440.

Engledow, F.L., Yule, G.U. The determination of the best value of the coupling-ratio from a given set of data. (1915) *Am. Naturalist* **49**: 127–128.

Fisher, R.A. On an absolute criterion for fitting frequency curves. (1912) *Mess. Math.* **41**: 155–160.

Fisher, R.A. On the 'probable error' of a coefficient of correlation deduced from a small sample. (1921) *Metron* **1**: 3–32.

Fisher, R.A. On the mathematical foundations of theoretical statistics. (1922a) *Phil. Trans. R. Soc.* **A 222**: 309–368.

Fisher, R.A. The systematic location of genes by means of crossover observations. (1922b) *Am. Naturalist* **56:** 406–411.

Fisher, R.A. *Statistical Methods for Research Workers.* (1925) Oliver and Boyd, Edinburgh.

Fisher, R.A. *Statistical Methods for Research Workers.* 2nd edition. (1928). Oliver and Boyd, Edinburgh.

Fisher, R.A., Balmukand, B. The estimation of linkage from the offspring of selfed heterozygotes. (1928) *J. Genet.* **20:** 79–92.

Fisher, R.A. The amount of information supplied by records of families as a function of the linkage in the population sampled. (1934). *Ann. Eugen.* **6:** 66–70.

Fisher, R.A. The detection of linkage with 'dominant' abnormalities. (1935). *Ann. Eugen.* **6:** 187–201.

Haldane, J.B.S. The probable errors of calculated linkage values, and the most accurate method of determining gametic from certain zygotic series. (1919a). *J. Genet.* **8:** 291–297.

Haldane, J.B.S. The combination of linkage values, and the calculation of distances between the loci of linked factors. (1919b) *J. Genet.* **8:** 299–309.

Haldane, J.B.S. Methods for the detection of autosomal linkage in man. (1934) *Ann. Eugen.* **6:** 26–65.

Harris, J.A. A simple test of the goodness of fit of Mendelian ratios. (1912) *Am. Naturalist* **46:** 741–745.

Morgan, T.H. Random segregation versus coupling in Mendelian inheritance. (1911) *Science* (n.s.) **34:** 384.

Muller, H.J. The mechanism of crossing-over. (1916a) *Am. Naturalist* **50:** 193–221.

Muller, H.J. The mechanism of crossing-over II. (1916b) *Am. Naturalist* **50:** 284–305.

Muller, H.J. The mechanism of crossing-over III. (1916c) *Am. Naturalist* **50:** 350–366.

Muller, H.J. The mechanism of crossing-over IV. (1916d) *Am. Naturalist* **50:** 421–434.

Pearson, K. On the criterion that a given system of deviations from the probable in the case of a correlated system of variables is such that it can be reasonably supposed to have arisen from random sampling. (1900) *London, Edinburgh, and Dublin Philosoph. Mag.* Fifth series **50:** 157–175.

Yule, G.U. On the association of attributes in statistics: with illustrations from the material of the Childhood Society, & co (1900) *Phil. Trans. R. Soc.* A **194:** 257–319.

Yule, G.U., Engledow, F.L. The determination of the best value of the coupling ratio from a given set of data. (Abstract) (1914) *Cam. Univ. Reporter* **44:** 757.

SECTION II

LINKAGE IN DOMINANT, RECESSIVE AND OLIGOGENIC DISEASE

Chapter 3
Oligogenic linkage and map integration

Newton E. Morton and Pietro Lio

◆

Mapping spans three genetic classes (*Table 3.1*). Major genes have high penetrance. If rare they may account for little of the liability variance, but they have large effects in segregating families. After 25 years of methodological development, lods (logarithms of odds) were established as the method of choice (Morton, 1955). After another 25 years, the number of polymerase chain reaction (PCR)-based markers became large enough so that mapping became easy, except for 'nonspecific' diseases like cognitive and sensory defect, where the number of contributing major loci is high. Oligogenes (called leading factors by Sewall Wright, 1968) are genes with smaller effects in segregating families, but still large enough (unless rare) to be detected by linkage. Polygenes are the remaining class, with effects so small that linkage has little power to detect them, but allelic association at candidate loci is sometimes successful (Risch and Merikangas, 1996).

Table 3.1. Phenotypic classes of alleles

Allelic class	Segregation analysis	Gene frequency	Effect on liability	Displacement t	Effect on penetrance	β
Major gene	+	Rare	Megaphenic	> 1.5	High	> 1.5
Oligogene	0	Common	Mesophenic	0.5–1.5	Low	0.1–1.5
Polygene	0	Common	Microphenic	< 0.5	Very low	< 0.1

GENETIC MAPPING OF DISEASE GENES
ISBN 0-12-232735-7

Oligogenic linkage

After 55 years of linkage tests based on moments and correlations, reminiscent of the earlier history with major loci, lods (either parametric or nonparametric) were recognised as the best method for oligogenic linkage. Parametric models specify dominance, gene frequency, displacement, and interactions for one or more contributory loci and therefore can allow for incomplete ascertainment. Nonparametric models confound these genetic variables (and so cannot allow for ascertainment), but of course they are not model-free. The β model assumes that the probability of an observation f on a pair of relatives who share k genes identical by descent (IBD) at a candidate locus is proportional to $e^{\beta k}$ (Morton, 1996). If the relatives are both affected f = 1, and so it is reasonable to take f for a quantitative trait as $X_1 X_2$, where X_1, X_2 are values defined on the individuals as deviations from the population mean and U is the expected value of $X_1 X_2$ for pairs of affected relatives, or approximately μ^2 if μ is the mean of X for an affected individual. Extreme discordant pairs have large negative values of f, extreme concordant pairs have large positive values, and pairs near the mean have values of f near zero. In the β model they are given little weight, unlike the Haseman–Elston method in which a small squared difference can occur anywhere in the phenotypic distribution. The β model has been shown to be more powerful than other nonparametric tests (Collins *et al.*, 1996).

Recently we have extended the β model to multipoint analysis, using the MAPMAKER/SIBS platform to compute the inheritance vector from hidden Markov chains (Krugylak and Lander, 1995). The two parameters are β and S, the location in centiMorgans (cM) on a marker map. Ideally this would be sex-specific (S_m and S_f). We have applied β to data on insulin-dependent diabetes (IDDM) kindly provided by John Todd (Davies *et al.*, 1994). In single locus tests the same lod is obtained as in parametric analysis (using the program COMDS which includes two loci, recombination, allelic association, and ascertainment). However, this identity does not hold for pedigrees more complicated than affected pairs or if the parametric program is incapable of maximum-likelihood estimation of dominance, gene frequency and penetrance under incomplete ascertainment (like LINKAGE and GENEHUNTER). Multilocus tests are substantially more powerful, with the β model retaining its superiority over the Δ model of MAPMAKER/SIBS and the nonparametric linkage model of GENEHUNTER (Lio and Morton, 1997). The estimate of β may be negative because the model is not constrained to the 'possible triangle', whereas constraint of the other models to the 'possible triangle' make the mean lod nonnegative, whether or not there is linkage (Holmans, 1993). Such biased tests are unsuitable for meta-analysis of multiple samples.

With an appropriate significance level, tests for oligogenic linkage have now reached the efficiency and reliability that lods provided for major loci 40 years ago.

Map integration

Map integration combines genetic and physical data by objective algorithms. The result is a summary map that assigns to each locus a vector of coordinates: band on the cytogenetic scale, megabases (ultimately base pairs) on the physical scale, and centimorgans on the sex-specific genetic scales. Radiation hybrids are sometimes presented on their own centiray (cR) scale, but this is so dependent on conditions of the experiment that it is of little use unless converted to an approximate physical scale. Physical, genetic and radiation hybrid maps take the short arm telomere (pter) as the origin. Partial maps take the locus closest to pter as the origin.

From 1990 to 1995 algorithms for map integration were being tested and the result was an integrated map (Collins *et al.*, 1995). Now the number of mapped loci is so great that recent attempts at publication in a scientific journal have been illegible (Dib *et al.*, 1996).

Despite dramatic improvement in the resolution and reliability of maps, there are serious limitations. The last attempt at an International System for Human Gene Nomenclature (ISGN) was in 1987. Changing data since then have not been reflected in a consensus about nomenclature, with the result that every database uses parochial symbols for which aliases must be found in any effort at map integration. Resources have been devoted to massive international databases for sequences and mutations, in each case with data presentation as the main objective, but the Genome Database (GDB) presents no data and is harshly criticised by the mapping community (van Heyningen, 1996). The GDB is mirrored at several national sites, which are so occupied with copying files that little effort is expended on creative work. Journals impose no standards in presentation of map data and so retrieval into a database is inordinately time-consuming. The number of mapped sequences will reach one million by the year 2005, when sequencing the human genome is expected to be complete. The role of maps in the sequencing effect is controversial (Venter *et al.*, 1996), but increasingly refined maps are indispensable for localization of disease genes, most of which are of unknown structure and function until they are positionally cloned.

The need for better maps becomes more critical as the most striking major loci are cloned. The remaining major genes are rarer and may have less obvious candidates. Complex inheritance adds the additional complications of low penetrance and high heterogeneity. Allelic association (linkage disequilibrium) can be

19

a powerful tool in a sufficiently small region, but not in a larger region or with a genome scan unless the density of markers is high, since allelic association is rarely detectable beyond 1 cM even for major genes with larger effect, short history, and less heterogeneity. Positional cloning for disease genes under complex inheritance cannot succeed unless map quality is improved. Linkage cannot determine order within 1 cM except in enormous samples, and so integration of data from physical and radiation hybrid maps is essential. Radiation hybrids provide the currently best way to assign an approximate location to expressed sequence tags (ESTs), which are often monomorphic.

Mapping resources

The mapping resources mentioned in this paper are available through the Internet (Collins and Morton, 1996). The location database (ldb) is currently accessed through the World Wide Web (WWW) at `http://cedar.genetics.soton.ac.uk/public_html/`. It consists of a directory for each human chromosome (as chrom *n* for chromosome number $n = 1, \ldots 22$, X, Y). Within each directory there are subdirectories for summary, genetic, physical, clonal, and radiation hybrid data and corresponding partial maps. Summary maps are also available on CD-ROM (Genome Interactive Databases, John Libby Eurotext).

Interacting with ldb, the pairwise mapping program MAP creates linkage, radiation hybrid, and nondisjunctional maps from corresponding data. Interference and error filtration are allowed for by parameters p and c, respectively. Estimates of these parameters are consistent among chromosomes. Efforts are now being made to create an international mapping database that would include locations as well as other locus-oriented data.

References

Collins, A., Forabosco, P., Lawrence, S., Morton, N.E. An integrated map of chromosome 9. (1995) *Ann. Hum. Genet.* **59 (Pt 4):** 393–402.

Collins, A., MacLean, C.J., Morton, N.E. Trials of the β model for complex inheritance. (1996) *Proc. Nat. Acad. Sci. USA* **93:** 9177–9181.

Collins, A., Morton, N.E. (1996) Human genome mapping. In Adolph, K.W. (ed.) *Human Genome Methods.* CRC Press Inc, Cleveland.

Davies, J.L., Kawaguchi, Y., Bennett, S.T., Copeman, J.B., Cordell, H.J., Pritchard, L.E., Reed, P.W., Gough, S.C., Jenkins, S.C., Palmer, S.M., Balfour, K.M., Rowe, B.R., Farrall, M., Barnett, A.H., Bain, S.C., Todd, J.A. A genome-wide search for human type 1 diabetes susceptibility genes. (See comments.) (1994) *Nature* **371 (6493):** 130–136.

Dib, C., Faure, S., Fizames, C., Samson, D., Drouot, N., Vignal, A., Millasseau, P., Marc, S., Hazan, J., Seboun, E., Lathrop, M., Gyapay, G., Morissette, J., Weissenbach, J. A comprehensive genetic map of the human genome based on 5,264 microsatellites. (1996) *Nature* **380 (6570):** 152–154.

Holmans, P. Asymptotic properties of affected-sib-pair linkage analysis. (1993) *Am. J. Hum. Genet.* **52 (2):** 362–374.

Kruglyak, L., Lander, E.S. Complete multipoint sib-pair analysis of qualitative and quantitative traits. (1995) *Am. J. Hum. Genet.* **57 (2):** 439–454.

Lio, P., Morton, N.E. Comparison of parametric and nonparametric methods to map oligogenes by linkage. (1997) *Proc. Nat. Acad. Sci. USA* (In press).

Morton, N.E. Sequential tests for the detection of linkage. (1955) *Am. J. Hum. Genet.* **7:** 277–318.

Morton, N.E. Logarithm of odds (lods) for linkage in complex inheritance. (1996) *Proc. Nat. Acad. Sci. USA* **93:** 3471–3476.

Risch, N., Merikangas, K. The future of genetic studies of complex human diseases. (1996) *Science* **273:** 1516–1517.

van Heyningen, V. Changing tack on the map. (1996) *Nat. Genet.* **13 (2):** 134–137.

Venter, J.C., Smith, H.O., Hood, L. A new strategy for genome sequencing. (1996) *Nature* **381 (6581):** 364–366.

Wright, S. Evolution and the Genetics of Populations, Vol 1. *Genetic and Biometric Foundations.* (1968) University of Chicago Press, Chicago.

Chapter 4
Genetic mapping of complex disorders
Jurg Ott

Introduction

In today's vocabulary of human geneticists, mapping a disease means localizing on the human gene map the gene (or one of the genes) that causes the disease, that is, assigning the gene's location on a chromosome. This has been achieved for many monogenic traits (i.e. traits following a mendelian mode of inheritance). The next step in this so-called positional cloning approach is to characterize and sequence the gene and to elucidate its function.

Many of the common diseases seem to have a genetic component, but their mode of inheritance is unclear. Examples of such so-called complex traits are schizophrenia, some cancers and diabetes mellitus. They are generally thought to be under the influence of multiple environmental and genetic factors. Presumably, involvement of several genes is the rule rather than the exception. Current ideas on the genetic analysis of complex traits and methods for localizing genes implicated in these traits are outlined below. It is assumed that the reader has a basic understanding of genetic linkage analysis. Otherwise, introductory textbooks on this topic should be consulted (Ott, 1991; Terwilliger and Ott, 1994). An overview of analysis methods for complex traits can also be found in a paper by Lander and Schork (1994).

Genetic basis of traits

The genetic basis of a trait may be established by demonstrating that the trait runs in families. Of course, common family environmental effects and infection

GENETIC MAPPING OF DISEASE GENES
ISBN 0-12-232735-7

may also lead to familial aggregation of affected individuals, but barring such factors, familial occurrence is generally interpreted as evidence for a genetic effect. The extent to which a trait runs in families may be measured by the ratio of the recurrence risk in monozygotic twins versus that in dizygotic twins, or the ratio of recurrence risk in a sibling versus population incidence (Plomin *et al.*, 1994).

Genetic effects may be the consequence of a single gene (mendelian mode of inheritance), a small number of genes (oligogenic inheritance), or a large number of genes with a small effect each (polygenic inheritance, for example body height, skin color). Distinguishing between these effects is often difficult. Segregation analysis or a newer method described by Risch (1990) may be carried out to ascertain whether single genes are involved in a trait. Generally, however, once good evidence for heritability of a trait has been established, researchers proceed directly to genetic linkage analysis to localize disease genes. Presumably, many complex traits are due to a rather small number (less than a dozen) of genes of varying effects on disease susceptibility. Depending on the power of the analysis and the sample size (number of families investigated), the few most influential genes will be detected, as for example in diabetes mellitus (Davies *et al.*, 1994).

Parametric approaches

When multiple genes are involved in a trait, the question is how these genes interact to lead to disease susceptibility. One possible inheritance model is as follows: each of n disease loci has two alleles, N and D, where the D alleles are disease alleles. An individual is only then susceptible to the disease when he or she carries a minimum of t ($\leq 2n$) D alleles irrespective of which loci these alleles come from. For complex traits, the number of disease loci and the mode of interaction among them is generally unknown. For this and practical reasons, linkage analysis between a trait and a marker is almost always carried out under an assumed single-locus (mendelian) mode of disease transmission. Analyzing data under a clearly 'wrong' model is expected to result in a loss of power and in a biased estimation of the recombination fraction. For some two-locus modes of inheritance, these negative effects have been shown to vary depending on the mode of inheritance; in some cases, loss of informativeness is small. Also, allowing for heterogeneity on a routine basis tends to reduce the bias in the recombination fraction estimate (Schork *et al.*, 1993).

In linkage analysis between one of the disease loci and a neighboring genetic marker, the recombination fraction tends to be overestimated. For this reason, in multipoint analysis, a disease gene is typically estimated away from each marker, that is, off the marker map. Therefore, for complex traits, the linkage method of choice is two-point linkage analysis, and multipoint analysis should be avoided, at least initially.

In many mendelian and complex traits, penetrance is incomplete, that is, a susceptible individual carrying a disease-predisposing genotype has a less than 100% chance of developing the disease. (The reverse case, that is, a nonsusceptible individual developing the disease, is called a phenocopy.) Under many inheritance models, affected individuals carry more information on linkage than unaffected individuals. For this reason, attention is often focussed on affected individuals only, while information from unaffected individuals is ignored. In practice, the easiest way to achieve this is to divide all penetrances by a large number such as 1000; this does not change the lod score, but renders all unaffected individuals virtually 'unknown' (Terwilliger and Ott, 1994).

When the mode of inheritance is unknown and analysis is most likely carried out under unrealistic assumptions, what inheritance model should be chosen for the trait to be mapped? Until recently, it was recommended that models be as realistic as possible. However, as will be seen in the next section, it is probably not too important what specific penetrance values are chosen. For example, 0.005 for phenocopies and 0.750 for genetic cases seem suitable for many diseases; these two values define a penetrance ratio (risk ratio) (Ott, 1994) of 150, which specifies a moderately strong genetic effect. What matters more is whether analysis is carried out under a dominant or recessive mode of inheritance. Therefore, two analyses are usually carried out, one under dominant and one under recessive inheritance.

For many complex traits, various diagnostic schemes exist ranging from severely affected states to mild disease symptoms. It is generally unclear which diagnostic scheme is genetically relevant, that is whether a hypothesized disease gene has a narrow spectrum of action causing only well-defined 'core' illness or whether it has a wide spectrum leading to many different manifestations, some of which may only loosely be connected with the disease. Because of this uncertainty, multiple analyses are typically carried out under different diagnostic schemes. In effect, this strategy amounts to maximizing the lod score over diagnostic schemes (and perhaps genetic analysis models). Consequently, there is an increased chance of a false positive result (type I error). Using computer simulation, the actual type I error may be assessed (Weeks *et al.*, 1990). When multiple analysis schemes are contemplated, it is important that they be defined and fixed before analysis starts. Otherwise, a rigorous assessment of the false positive rate becomes difficult, even with computer simulation, and results can no longer be meaningfully interpreted.

Nonparametric linkage analysis

The lod score method referred to above is called parametric because parameters of disease inheritance are specified (perhaps estimated) in the analysis. In contrast to it, nonparametric methods of linkage analysis exist in which no reference is made to the mode of inheritance of the disease; all analysis is done on the basis of marker inheritance. Connection to disease is established in that attention is focussed on affected individuals only. These methods are often said to be independent of the mode of disease inheritance. However, such a statement is misleading: although the mode of inheritance is disregarded in the analysis, properties (e.g. power) of the various nonparametric methods depend on the mode of inheritance of the disease. Also, while users of nonparametric methods need not specify a mode of inheritance, they still face the problem of having to decide who is affected and unaffected.

A well-known nonparametric linkage analysis method is the Affected Sib Pair (ASP) method: finding that affected siblings share more marker alleles as copies of parental alleles, that is, identical by descent (IBD), than expected by chance indicates the presence of a susceptibility gene in the vicinity of the marker in question. For example, assume that the mother's genotype is 1/2 and that of the father is 3/4. If two affected siblings have genotypes 1/3 and 1/4 they share one allele (i.e. allele 1) IBD. By chance alone, the proportions of ASPs sharing 0, 1 and 2 alleles IBD are expected to be 0.25, 0.50 and 0.25, respectively.

On a per parent basis, IBD sharing in ASPs is either 0 or 1. As a test for linkage, the observed proportion of alleles shared is then compared with the expected proportion (under no linkage) of 0.50. Under certain conditions, this test is equivalent to the lod score test in the sense that critical values of the test statistics exist such that when one test is significant so is the other (Hyer *et al.*, 1991; Knapp *et al.*, 1994). These conditions refer to the lod score analysis and are as follows: only nuclear families are analyzed (large pedigrees broken down into nuclear families), parental phenotypes are disregarded ('unknown'), analysis is under a recessive mode of inheritance with full penetrance and no phenocopies (Knapp *et al.*, 1994).

This equivalence demonstrates that ASP tests must be most powerful for recessively inherited traits. For dominant traits, only one of the parents is informative for linkage so that roughly twice as many ASPs are required for the same power to map a dominant versus a recessive gene.

At this point, consider a question raised in the Parametric Approaches section: what penetrance values should be chosen in the single-locus model for analysis of a complex trait? The equivalence between ASP and lod score analysis suggests penetrances of 0 (for phenocopies) and 1 (for genetic cases), that is an infinite penetrance ratio. In practice, researchers want to allow for some occurrence of phenocopies. Therefore, reasonable choices for penetrances correspond to high penetrance ratios.

If more than two affected siblings occur in a sibship, various solutions have been proposed, for example, to form all possible pairs, but reduce the weight of each pair in an appropriate fashion. The simplest solution, however, is to emulate the ASP method by a lod score analysis in which occurrence of multiple affected siblings is no problem.

Various extensions of ASP methods have been introduced. For example, quantitative traits may be analyzed by the Haseman–Elston method (Haseman and Elston, 1972; Wilson and Elston, 1993), which has been implemented in one of the SAGE (Statistical Analysis in Genetic Epidemiology) programs. Also, although multipoint analysis by the lod score method is not recommended for complex traits (see above), multipoint ASP methods focus on marker inheritance only (Ott, 1996) and represent efficient ways of nonparametric linkage analysis (Kruglyak and Lander, 1995).

Genome-wide linkage analysis

In addition to multiple classification schemes tried in linkage analysis, testing for linkage with many markers also represents a multiple testing problem. If n independent tests are carried out and each individual test is called significant when the resulting significance level is below the value, p, then the 'global' significance level (the probability that at least one of the tests turns out significant just by chance) is given by $1 - (1 - p)^n$. For example, since a critical lod score of 3 asymptotically corresponds to a one-sided p value of 0.0001 (Ott, 1991), testing linkage with $n = 100$ unlinked markers leads to a genome-wide significance level of 0.01, which approximately corresponds to a critical lod score of $3 - \log_{10}(100) = 1$ (Kidd and Ott, 1984).

It is interesting to see how different generations of linkage analysts have dealt with the multiple comparison problem of marker testing. Originally, because of the paucity of markers, no problem existed. The stringent criterion of a critical lod score of 3 was proposed to have high power in the presence of a low prior probability (approximately 5%) of linkage (Morton, 1955). Subsequently, with multiple markers to be tested, it was recognized that eliminating portions of the genome from consideration increases the prior probability of linkage for the remainder of the genome. These two effects (multiple tests, increased prior probability of linkage) approximately cancel each other, which furnished justification for retaining the lod = 3 criterion, but only when a disease locus was known to exist somewhere in the genome. In the case of complex traits and when linkage was used to prove existence of disease genes, there was no guarantee that a locus would eventually be found. Consequently, there was no increased prior probability of linkage and no justification for not increasing the critical lod score in the face of multiple testing. Therefore, much higher critical lod scores for complex traits were discussed.

This line of thinking was in the framework of prior and posterior probabilities. In the framework of likelihood ratio tests, which is the predominant current approach in human linkage analysis, linkage analysts want to see which lod score threshold leads to what probability of a false positive result (i.e. significance level), which is a significant result without there being linkage. In other words, they never even invoke linkage; all calculations on significance levels are carried out under the assumption of no linkage, in which case it is immaterial whether one works with a mendelian or complex trait. Lander and Kruglyak (1995) have approximately shown which critical locus-specific lod score corresponds to a genome-wide significance level of 5%. Depending on the type of linkage analysis, their theoretical results postulate locus-specific lod score thresholds between 3.3 and 3.8. Previously, based on computer simulation of an oligogenic threshold trait and a critical lod score of 4, Suarez *et al.* (1994) calculated a genome-wide (106 unlinked markers) significance level slightly exceeding 5%.

To make a long story short: the classical lod score threshold of 3 (or a value somewhat higher than this) is considered appropriate for genome-wide linkage analyses, at least theoretically. To evaluate the significance of an observed maximum lod score in a genome screen, the best approach is to estimate the empirical significance level (the probability that the observed maximum lod score is attained or exceeded just by chance) by computer simulation, for example, with the aid of the SIMULATE program (Terwilliger *et al.*, 1993). It may be obtained from the anonymous file transfer protocol (FTP) site, `linkage.rockefeller.edu`

In psychiatric genetics, there are well-known reports of maximum lod scores way above 3 that could not be replicated in subsequent studies and seem to have been false positive results. They led to the recommendation that one should 'be skeptical about reports of psychiatric disease linkages at lod scores of less than 6' (Robertson, 1989). How can one explain the discrepancy between the linkage results in psychiatric traits and the genome-wide critical lod scores of less than 4 discussed above? The following three points come to mind.

- A genome-wide critical lod score of slightly more than 3 corresponds to a significance level of 5% (i.e. five out of 100 significant results are expected to be false) and the problematic linkage reports of psychiatric illnesses may happen to fall into this category.

- The recommended critical lod scores do not allow for multiple testing in the form of multiple data analyses under different disease classification schemes.
- Positive linkage results are more likely to be published than negative results.

Acknowledgement

This work was supported by grant HG00008 from the US National Center for Human Genome Research.

References

Davies, J.L., Kawaguchi, Y., Bennett, S.T., Copeman, J.B., Cordell, H.J., Pritchard, L.E., Reed, P.W., Gough, S.C., Jenkins, S.C., Palmer, S.M., Balfaur, K.M., Rowe, B.R., Farrall, M., Barnett, A.H., Bain, S.C., Todd, J.A. A genome-wide search for human type 1 diabetes susceptibility genes. (1994) *Nature* **371 (6493):** 130–136.

Haseman, J.K., Elston, R.C. The investigation of linkage between a quantitative trait and a marker locus. (1972) *Behav. Genet.* **2 (1):** 3–19.

Hyer, R.N., Julier, C., Buckley, J.D., Trucco, M., Rotter, J., Spielman, R., Barnett, A., Bain, S., Boitard, C., Deschamps, I., *et al.* High-resolution linkage mapping for susceptibility genes in human polygenic disease: insulin-dependent diabetes mellitus and chromosome 11q. (1991) *Am. J. Hum. Genet.* **48 (2):** 243–257.

Kidd, K.K. and Ott, J. (1984) Power and sample size in linkage studies. *Cytogenet. Cell Genet.* **37:** 510–511.

Knapp, M., Seuchter, S.A., Bäur, M.P. Linkage analysis in nuclear families. 2: Relationship between affected sib-pair tests and lod score analysis. (1994) *Hum. Hered.* **44 (1):** 44–51.

Kruglyak, L., Lander, E.S. Complete multipoint sib-pair analysis of qualitative and quantitative traits. (1995) *Am. J. Hum. Genet.* **57 (2):** 439–454.

Lander, E., Kruglyak, L. Genetic dissection of complex traits: guidelines for interpreting and reporting linkage results. (1995) *Nat. Genet.* **11 (3):** 241–247.

Lander, E.S., Schork, N.J. Genetic dissection of complex traits. (1994) *Science* **265 (5181):** 2037–2048.

Morton, N.E. Sequential tests for the detection of linkage. (1955) *Am. J. Hum. Genet.* **7:** 277–318.

Ott, J. *Analysis of Human Genetic Linkage.* 2nd edn (1991) Johns Hopkins University Press, Baltimore.

Ott, J. Choice of genetic models for linkage analysis of psychiatric traits. In Gershon, G.S., Cloninger, C.R. (eds) *Genetic Approaches to Mental Disorders.* pp. 63–75. (1994) American Psychiatric Press, Washington.

Ott, J. Complex traits on the map. (1996) *Nature* **379 (6568):** 772–773.

Plomin, R., Owen, M.J., McGuffin, P. The genetic basis of complex human behaviors. (1994) *Science* **264 (5166):** 1733–1739.

29

Risch, N. Linkage strategies for genetically complex traits. II. The power of affected relative pairs. (1990) *Am. J. Hum. Genet.* **46 (2):** 229–241.

Robertson, M. False start on manic depression. (News; comment.) (See comments.) (1989) *Nature* **342 (6247):** 222.

Schork, N.J., Boehnke, M., Terwilliger, J.D., Ott, J. Two-trait-locus linkage analysis: a powerful strategy for mapping complex genetic traits. (1993) *Am. J. Hum. Genet.* **53 (5):** 1127–1136.

Suarez, B.K., Hampe, C.L., Van Eerdewegh, P. Problems of replicating linkage claims in psychiatry. In Gershon, G.S., Cloninger, C.R. (eds) *Genetic Approaches to Mental Disorders.* pp. 23–46. (1994) American Psychiatric Press, Washington.

Terwilliger, J.D., Speer, M., Ott, J. Chromosome-based method for rapid computer simulation in human genetic linkage analysis. (1993) *Genet. Epidemiol.* **10 (4):** 217–224.

Terwilliger, J.D., Ott, J. *Handbook of Human Genetic Linkage.* (1994) Johns Hopkins University Press, Baltimore,

Weeks, D.E., Lehner, T., Squires Wheeler, E., Kaufmann, C., Ott, J. Measuring the inflation of the lod score due to its maximization over model parameter values in human linkage analysis. (1990) *Genet. Epidemiol.* **7 (4):** 237–243.

Wilson, A.F., Elston, R.C. Statistical validity of the Haseman–Elston sib-pair test in small samples. (1993) *Genet. Epidemiol.* **10 (6):** 593–598.

Chapter 5
Recessive disease and allelic association
J.H. Edwards

Recessive disorders of the eye are common in childhood and their study has the advantages that patients are usually accompanied by their mother and any affected sibs usually attend the same hospital. The mutations can be assumed to be old, and, if at the same locus, most are likely to be identical. However, compared to many other disorders, there are the major diagnostic disadvantages that biopsy is impractical, generalised metabolic disturbance unusual and the phenotypes related to different loci can rarely be distinguished with certainty.

The analysis by LINKAGE and related methods was highly efficient in the decade dominated by the restriction fragment variants with fewer marker loci, most with only two alleles, and has provided most of the established linkages. The necessary assumptions of a single disease locus, accurately estimated allelic proportions, sufficient distance between the marker loci for allelic association to be unusual, and all mutations to be different, did not lead to errors commensurate with the increased efficiency available from 'splinting' neighbouring sets of loci, most of which were uninformative in most sibships. I will discuss deductive approaches which lose both information and misinformation, but have the advantages of simplicity and simple graphical representation.

Few of the assumptions necessary for inferring linkages are realised in contemporary studies of recessive disorders and, in eye disorders, even the assumption of a single locus is unlikely to be realised except in populations carefully selected for isolation. Even when the null hypothesis of no heterogeneity is realistic the confounding of a recombinant event from a second locus in data from

GENETIC MAPPING OF DISEASE GENES
ISBN 0-12-232735-7

multiple small families deprives tests of much power. In the limiting and otherwise highly efficient case of affected sib pairs in families with only two affected individuals there is no information on heterogeneity. The advantages of working with isolated populations are that a disorder common in one may be absent or rare in another and the common ancestor responsible is likely to be relatively recent. The smaller the population, and especially the smaller its founder group, the smaller the number of different alleles with recessive manifestation. However, laboratory expertise is rarely established in the most appropriate populations at an adequate level to exploit this. Finland, Norway and Iceland are among the distinguished exceptions, but their major and disproportionate contributions have not been dominated by eye disease.

Although established methods of linkage analysis are readily available and provide a rough screen, they lack power when more than one locus is present and are liable to error by defining a confidence range that excludes the main locus. Their major shortcoming is that they do not exploit the between-family information on linkage. While the analysis of within-family linkage is compromised by unrealistic assumptions, between-family analysis is compromised by different assumptions and has different shortcomings. It is more powerful in detecting close linkage when it requires far smaller investments in both clinical and laboratory resources, but lacks power for more distant linkages. How close is close is considered below.

All linkage studies are based on allelic associations in relatives, known or unknown. If loci are linked then alleles are associated, although the obverse is not necessarily true. The extent of association depends on the closeness of the loci in space and the coancestry in time. Standard linkage methods are based on the close relatives who comprise what is usually termed a family. However, a species is merely a very large family with all members variously, mostly unknowingly, related. The term 'large family' in conventional usage covers overlapping sets of relatives known to each other. Subsets, varying from the large families of arbitrary structure sometimes available, the full three generation set of a couple with their parents and children, the nuclear family of parents and children, sib pairs with or without their parents, parent–child pairs and the distantly and unknowingly related individuals who form populations can all provide both information and misinformation.

In recessive disorders the problem of within-family analysis is simplified by the bulk of the information being available from small families with affected children. For simplicity I will consider only subsets of one, two and four individuals and sufficient allelic information for any allele to have its parentage defined if a parent has been typed. The power of the procedure varies virtually pro rata with heterozygosity for the multiallelic loci in current use.

Allelic association between relatives declines with distance, which is defined by the number of connecting meioses. Two individuals g generations from a common ancestor are $2m$ meioses distant. Although with large numbers of generations the number through each parent will usually differ, it is simpler to assume symmetry. This makes little difference, and in any case only the vaguest estimate of the length of either line is possible. There is no fundamental difference between regular families, isolated pairs of relatives, sets of identical pairs, including sib pairs with or without parents, and populations of individuals related in unknown ways. The latter, the ultimate family, consists of large numbers of 'families of one' connected by unknown numbers of meioses. Inferences may be compromised by allelic associations unrelated to linkage due to poor miscegenation following recent immigration. This potential cause of spurious linkage has led to the development of tests for allelic association within families, the so-called transmission equilibrium tests. These compare the allele proportions in the parental alleles transmitted and not-transmitted. In fact transmission is not at issue, since this is equal in the population, but ascertainment for affected children will lead to an excess of alleles associated with the disease compared to those not received from a parent. This is a procedure for comparing allele proportions in parents and children which exploits the bias of selecting parents who have affected children. It can also be achieved through parent–child pairs with higher efficiency and no bias for familiarity and will be considered below under allelic association. In mendelian, as opposed to multifactorial, disorders the bias from familiarity should hardly arise due to the high penetrance.

In practice the problem of false trails through poor miscegenation, which within family analyses will reduce, seems small except where there has been substantial immigration in the last few centuries and these immigrant groups are not easily defined by name, language, religion or appearance. In most of Europe the adequacy of assuming good miscegenation on easily defined populations is supported by both historical evidence and the extensive data, with numerous well-established associations related to linkage, from studies at the HLA locus (Tomlinson and Bodmer, 1995). Confusion requires the allele proportions at both loci to differ in the two populations.

As allelic association does not necessarily imply linkage the term has advantages over both the historically correct 'linkage inequality' of Robbins (Robbins, 1918), Fisher's 'linkage equilibrium' (Fisher, 1930), which assumed a stable inequality maintained by selection, or the more usual 'linkage disequilibrium'. None of these terms necessarily relates to linkage and the descriptive term allelic association has advantages (Edwards, 1980). Individuals with identical rare alleles will usually be more closely related than randomly defined individuals through a more recent common ancestor and, in recessive disorders, the offending allele can be assumed to be old. Some, such as cystic fibrosis, most thalassaemias and juvenile Batten's disease clearly preceded the Roman invasions and other pre-Christian migrations in Northern Europe, and most mutations will have preceded the massive migrations from continental Europe, Africa and India over the last three centuries. Although it is usual to assume mutants mutate and marker alleles do not, now that markers are selected for variability, which is obviously correlated with mutation rate, the marker mutation rate is likely to greatly exceed that of the mutant allele, both before and after mutation. Methods depending on large numbers of meioses cannot distinguish recombination from mutation. If closely linked loci lack allelic association, at least one must be highly mutable. With several very close loci the distinction should be evident on inspection. All alleles are mutants but it is convenient to restrict the term to the alleles at the offending loci capable of leading to mendelian, although not other, disorders. This conforms to standard usage in other species.

Robbins (1918) showed that the degree of association decays at $(1-t)$ per generation where t is the recombination fraction. Very approximately, a segment with a functional or genetic length of a centimorgan will have its continuity maintained with a half-life of about 70 generations $(0.99^{70} = 0.495)$ or about two thousand years. As the genetic length of the human genome is about 30 morgans or 3000 centimorgans and the structural or physical length of the haploid or gametic set is about 3000 million base pairs each megabase averages a functional length of about a centimorgan. That is, any segment of about a megabase will have a 'backward half-life' of about two thousand years. Any pair of gametes in any population of less than a few million inhabitants whose members can reasonably assume that most of their ancestors were from the same region one thousand years ago are likely to have numerous segments of several megabases in common whose identity has been undisturbed by recombination. If any locus is selected the chance of a common segment to one side is about $1 - (1-1/200)^g$ in g generations and continuity over both sides has a half-life of about 35 generations. Only the segment including the offending allele will have been ascertained and, depending on the distance of the coancestry and the number of neighbouring loci, may be sufficiently conspicuous against the background noise of diverse segments for any mathematical aids to credulity to be unnecessary. An example will be shown below. Similar arguments apply to other species: there are probably less than a hundred points of species interchange between man and mouse, and far fewer between man and either ungulates or carnivores. Any randomly defined locus is very unlikely not to be central to at least a megabase in another mammal. Sheep, cows, dogs and cats, as well as laboratory species, have the major advantage that absence of both locus and allelic heterogeneity can be assumed or imposed by breeding (see Chapter 17).

Standard methods of linkage analysis compare the observed pattern of manifestation in a family on the assumption that the allele proportions are known, and if several marker loci are analysed, that they are correctly ordered and positioned and not so close that allelelic association is sufficient to affect either analysis or map-making. All mutations are assumed different. These assumptions, when justified, allow increased efficiency over Morton's original formulation (Morton 1955), which is exact but excludes families from which information can be extracted if the allele proportions are known. Since almost all information on within-family linkage in recessive disease is in small families, exact analysis is simple from his original tables and, with multiallelic codominant loci little information would be lost. So far as I am aware these are not yet readily available for direct computer entry. A table allowing conversion from same and different proportions of marker alleles in sib pairs to recombination fractions and is given in the Appendix.

In the nuclear family of a couple with their children and fully informative markers there is no relevant ancillary information. All the information is conveyed by counts of allelic identity and difference in the informative alleles of those affected. If a single offending locus and marker is involved, Morton's tables will extract all the information, while simple significance tests will be the most powerful discriminant at a cost of not providing simultaneous estimates of linkage, which will usually be biased and too small if their acceptance is conditional on the value of the lod as this favours the helping hand of chance. Multilocus analysis has nothing to offer in the limiting case of very close linkage when the parents are different heterozygotes at marker loci. The consequences of erroneous assumptions increase as the number of alleles increase, since estimates of allelic proportions, especially of rare alleles, lack precision. The variation in allele proportions and their association in different populations limits any advantage of centralised data analysis.

The restricted coancestry of cousin marriages is an obvious source of enrichment, but has limitations as cousin marriage is rare in most communities and only slightly increased in common disorders. A few cases with overlapping segments of homozygosity for the same alleles will provide evidence of the disease locus being within the longest common segment, although precise localisation will require more distantly related parents. In some disorders the excess of sporadic cases, the low incidence of cousin marriage and the variety of identical segments containing mutation in the affected suggest that early embryonic lethals may be common in our species. These lethals have the effect of spacing births, but rarely to an extent that would impair fertility sufficiently to reduce the number of children reared successfully. As recessive embryonic lethals would selectively reduce the birth of inbred offspring they could be advantageous in a thinly spread species.

Sporadic cases are usually omitted from linkage studies on the erroneous grounds that they are necessarily less informative. They have a very low chance of being very much more informative than familial cases as, in a small minority, even basic cytogenetic studies will pinpoint the lesion. The product of this low chance and ultimate benefit will usually exceed that from systematic studies restricted to numerical linkage analysis. While cytogenetic studies can define a lesion at a resolution of a few base pairs linkage studies are limited, on average, to a resolution of several million base pairs. In recessive disorders in particular, as will be shown later, sporadic cases may be more efficient than familial cases in terms of information per unit cost. Sporadic cases are also more numerous. If a couple of carriers have two children most pairs of children will be expected to be healthy (9/16) and 1/16 to have both affected. Six will have one child affected, a ratio of 6:1 sporadic:familial. All one-child families will, of course, be sporadic.

In most European populations the proportion of sporadic cases will be at least three-quarters. Cytogenetic examination of one affected person per family should be obligatory in any linkage study for mendelian disorders, both on theoretical grounds and on the evidence of the consequent localisation of aniridia, retinoblastoma, neurofibromatosis, Duchenne muscular dystrophy, several forms of retinitis pigmentosa and many other conditions. These highly informative cases are more likely to be found in sporadic cases as some translocations impair fertility and many are new and absent in the parents.

Large families, so important in dominant and X-linked disease, are of limited value in recessive disorders, and, when available, little is lost if the more evasive connecting members are not typed, while much is gained by the reasonable assumption of a single allele being involved. In any large family most connecting members will be dead and attempts to trace 'statistical' or 'probable' haplotypes across gaps involve massive computing resources, and, in recessive disorders, the problem can usually be resolved by reasonable assumptions and approximations.

I will discuss only three subsets of a family, the affected sib pair with parents, the parent–child pair, either affected, and finally the affected individual alone. All families can be dissected into various pairs defined by relationship, phenotype and marker. The atomic unit of the nuclear family with two children provides the standard benchmark for studies of power and efficiency of sib pairs with parents. Morton tabulated information against recombination fraction (Morton, 1955) and Risch (1990) extended this to other pairs of affected relatives. Standardised units simplify both administration and analysis in systematic studies and, in mendelian disorders, there is little bias involved in assuming all sib pairs within a family can be treated independently when the alleles transmitted can be deduced. Routine referral of DNA from all cases, and, when easily available, from a parent or child, is practical through cell banks and, where these banks can adapt to the powers of the polymerase chain reaction (PCR), a far cheaper and more extensive service could be provided which could also accept DNA from regions in which limited resources, poor transport, erratic postal services or extremes of climate made sending cells impractical. Even blood spotted on blotting paper and air dried is sufficient for many purposes, while dried saliva is better than nothing.

By the power of a method of analysis is meant the ability to define the presence of a linkage while efficiency is the precision with which the locus with the offending allele can be placed in terms of centimorgans from its nearest neighbours. I will use the terms foursomes, twosomes and orphans respectively to cover affected sib pairs with parents, parent–child pairs with one affected, and individuals without known typed relatives. Although multiple affected sib pairs may be defined in families with more than two affected children, I will ignore these: in mendelian disorders they provide an amount of information approximately equal to the number of ways they can be paired, but provide no extra information on between-family linkage as this is limited by the number of parents.

In a paper on non-mendelian disorders Lander and Shork (1994) distinguished large families, sib pairs and association studies in populations as distinct quarters of their 'fourfold way'. In the three relating to human studies linkage analysis pursues the same ends by the same means, the detection of segments of chromosome overrepresented in those affected. In both large families and affected sib pairs the limitation is ignorance of the number of different mutations, of the number and proportions of loci with mutant alleles involved, of the proportions of alleles at the marker loci and of the extent of allelic association between nearby markers. In association studies the limitation is ignorance of exact coancestry with the associated advantages of a large but unknown number of connecting meioses.

Allelic association will differ markedly between mutant-bearing and other gametes at markers near offending loci. Limited information on distance or even order in the genetic maps will not compromise the reliability of detection. It is known to be effective and powerful in segments of about a megabase due to the extensive studies on HLA and disease, and will obviously be more powerful in recessive disease, especially in small populations, as the nearest common ancestor is likely to be relatively recent as ascertained for a locus homozygous for a rare allele.

Finding a distant linkage is of far less value than a close linkage since sequencing even one megabase can hardly be justified, and even if sequenced or the sequence is known, there is little prospect of defining a point mutation against the background of normal variability. No within-family study can anticipate such precision with realistic resources in data, marker number and the precision of diagnosis and allele identification and recording. If loci are a mere centimorgan apart 70 meioses will be as likely as not to lead to non-recombinant. A detailed genetic map of the genome is often regarded as necessary, but it adds little relevant information if the marker loci are sufficiently informative. Although a genetic as opposed to a physical map is of limited use, the study of haplotypes ascertained by a disease locus can provide information on order beyond the power of data limited by smaller, although known, numbers of meioses.

Theoretical considerations

I will only consider the detection of linkage. The expectation of locus hetero-geneity makes the luxury of simultaneous detection and estimation available from the standard likelihood approaches inappropriate in most recessive disorders, especially those involving eye, ear or brain. Sufficient is known of the mechanism of genetic disorders to assume locus heterogeneity is likely due both to most enzymes working in sequential metabolic pathways and most being compounded of products from more than one locus. It is convenient to consider the three sub-sets of foursomes, twosomes and orphans of 100 affected children in which the children may be paired as sibs into 50 sib pairs, or unpaired but accompanied by a parent, or without known relatives.

In parent–child pairs of twosomes, I will discuss the usual affected-child situation. In English the word 'child' can cover any age in relation to a parent and for the basic analysis it is immaterial which is affected. Here I only consider the commoner affected-child situation. Although most recessive disorders present during childhood, a substantial proportion develop slowly to present in later life when a 'child', however old, is more likely to be available than a parent of an affected person. Recessive disease in the elderly is difficult to recognise from family histories, but is probably common in the sense organs since many forms developing in childhood are known, and milder variants must occur. I will con-sider only the 'ideal' situation, which is when a parent is present and the parental origin of all marker loci can be deduced. For both parents of a sib pair this is only possible if they are not only heterozygotes, but different heterozygotes. If h is the proportion of heterozygotes and only two alleles are present one parent will be fully informative in $2h(1 - h) + h^2$ or $h(2 - h)$ or, since h cannot exceed a half, less than three-quarters. However, diallelic markers are now rare and, if accompanied by such uninformative neighbours, require multilocus lod-score methods for their efficient analysis.

Double heterozygotes for identical alleles provide information that can be inferred but not deduced, which necessarily confounds parental origins. There are advantages in ignoring it with multiallelic markers. The allele proportions usual in markers in current use, which are selected for multiallelism, assures at least three distinct alleles in the majority of couples. The expected proportion of informative gametes from these couples will be informative in a proportion very close to h, which averages about 0.8, a proportion increasing with advancing techniques. While a single multiallelic marker will usually be informative a segment including neighbouring markers will be more informative provided it is not disrupted by meiosis. The shorter the segment the less likely this will be while what a longer segment gains in specificity it will eventually lose by rarity. The most informative length is usually obvious on examination, but its prediction depends on information, which can only follow analysis.

Affected sib pairs with parents: foursomes

Consider a set of affected sib pairs with parents, in which each parent has a mutant allele at a locus with a recombination fraction t relative to a marker locus with a large number of rare alleles allowing the parental origin of all alleles to be deduced. We also assume, as is reasonable in recessive disorders, that there are no new mutations within any sibship: that is, all mutations were conveyed from the grandparents and in all sibships the same locus is involved. As both sibs are affected, the chance of the alleles from a parental marker locus differing by origin will be:

$$b = 2t(1-t) \tag{Eq. 5.1}$$

The proportion that are the same will be:

$$a = 1-2t(1-t) \tag{Eq. 5.2}$$

On the null hypothesis of no linkage ($t = \frac{1}{2}$) these will be equal and the difference in proportion of those with the same and different alleles will be:

$$(a-b) = 1-4t(1-t) = (1-2t)^2 \tag{Eq. 5.3}$$

If there are n sib pairs:

$$n(a-b)^2/(a+b) = n(a-b)^2 = n(1-2t)^4 \tag{Eq. 5.4}$$

will have an expected distribution on the null hypothesis equal to χ^2 for one degree of freedom. It is convenient to use its square-root, χ, which has a value of the same order as Morton's z when either is about 3.0, although formally incommensurate.

If phase were known, as in the children of second cousin marriages with

39

typed grandparents, the 'idealised sib pairs' of Lander and Kruglyak (Lander and Kruglyak, 1995), the value would be $2n(1-2t)^2$, with a value of χ of $\sqrt{(2n)}\,(1-2t)$. They refer to this as Z, which should not be confused with Morton's z, which Ott (Ott, 1974) terms Z.

For a one tail test, appropriate in mendelian disorders, values exceeding 3.1 have a chance expectation of less than 0.001. The use of χ, which has the additional advantage of being restricted, in practice, to the range of 0–10, while in multifactorial disorders it can be negative in a meaningful way in relation to loci with some alleles conferring resistance.

Figure 5.1 show the expected value of χ against the recombination fraction for 50, 30 and 20 sib pairs.

All the information on the detection of linkage is provided by the parents

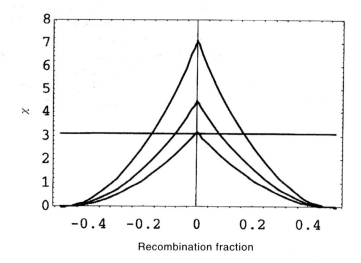

Figure 5.1 Values of χ against recombination fraction for 50, 30 and 20 foursomes with radii defined at a threshold of 3.1 ($p = 1/1000$). For 50 sib pairs the radius is 16 cM, for 30 sib pairs 7.7 cM and for 20 sib pairs zero.

transmitting different marker alleles with a mutant allele. For 100 affected children paired into 50 sibships with a very closely linked and fully informative marker the expected value of χ will be $\sqrt{50}$ or just over 7.0. For 30 and 20 children in 15 and 10 sibships, which are more realistic numbers in practice, it will be about 4 and 3 respectively. In practice it will be less due to markers not being fully informative, and reduced by the factor $(1-2t)^2$ for a recombination fraction of t as well as the additional approximate factor of the heterozygosity at the marker locus. It will only just be sufficient to be reasonably above the likely vagaries of chance on substantial numbers. The detection of even complete linkage at a probability of 0.001 for a single locus would usually require at least a dozen affected sib pairs. In any analysis the realised value of χ would, of course, be expected to be less than expected in half the analyses.

The value of t that is just sufficient to have an expectation of being significant at the level decided is known as the radius (Haldane and Carter, 1955). If at least one marker is always within this radius then there is a complete genome scan. We may visualise this as a rake with teeth separated by less than twice the radius of the offending locus. We assume that the rake will detect any locus it hits above a certain size rather than requiring it to be wedged between two teeth.

This is not an unreasonable model as if we envisage pebbles mass will vary as the cube of the radius. The more usual concept of a screen, riddle or sieve is simple if the marker loci are all equally spaced and equally variable, but holes of different sizes are difficult to manage in theory and may be impossible to display in practice. As the genome is linear and appropriately displayed by loci parallel to the rake a two-dimensional sieve is unnecessary. This 'sensitive' rake has the advantage of providing a compatible model with any number or spacing of teeth or even a single tooth. If the teeth are exactly one radius apart then the marker will occasionally miss hitting the offending locus, and will only wedge it half the time, since the actual radii will vary by chance about their expectation, and about half will be below it.

In mendelian disorders a full genome scan is now usually within range of practical numbers of affected individuals and markers. If all markers were equally informative the square root of the number of cases would determine the radius, while each marker defines a tooth. The product of the square of the number of sib pairs and the number of markers needs to be of the order of 100 000. The model conforms to normal methods of display as the rake can be drawn with the length of tooth proportional to some measure of evidence of linkage, and the evidence against linkage incorporated by standardising the total tooth length, as in the EXCLUDE program (Edwards, 1987). There is no null hypothesis since the offending locus must be somewhere and the logic is that of a horse race: the winner wins even though there is no guarantee of consistent winning in future.

The status of the 'runner up' is more difficult, but the possibility of a second locus should not be ignored. At the 1/1000 level of significance with 300 markers there would be a chance of $1-(1-1/1000)^{300}$ or 26% of an unlinked locus making the grade if all markers were unlinked. However about a twentieth are on the same chromosome and an 'equivalent number' of 2–4 per autosome, or about 50, would seem realistic. In which case the expectation of a false clue would be about 5%. It should be noted that due to allelic association even false clues will be supported by closely linked neighbours, although the ascertainment for recessive loci will lead to stronger 'solidarity' among the 'true' neighbours over longer segments. The runner up is obviously the second best bet as the long-term winner given no other information, and the difference between these two is relevant.

The addition of further markers on the same data after a preliminary screen is sometimes recommended, but it is both inefficient and logically unsound. As complete genome scans will often be impossible with practical numbers of patients and markers we may consider the problem of the significance to be accorded to the finding of the highest value of χ in one of n markers. The probability of at least as high a value due to chance is approximately np where p is the probability defined by the value of χ when np is small. The exact value is $1-(1-p)^n$. The prior probability of the locus being within range of a marker will be approximately nr/G where r is the radius and G the genome length. If this is of the order of 1/50 the usual 'rule of three' advanced by Morton will be a reasonable compromise between missing what is there and finding what is not. Whatever the largest value of χ substantial data deserve publication. The 'sniffing' approach of a preliminary screen followed by a denser screen on further data, with further marker loci concentrated around those markers showing suggestive evidence of linkage, has the disadvantage of often exiling the offending locus at an early stage clearing the way for the pretender.

Parent–child pairs, one affected: twosomes

A hundred affected individuals and a parent would, if fully informative to the extent that the child had an allele absent from the parent tested, provide two affected haplotypes from the child and one 'control haplotype' from the parent. I will discuss the affected-child problem: the affected adult can be handled in a similar way. The laboratory work would be halved while the clinical load would be severalfold less than in finding affected pairs and arranging for both parents and the other sib to be examined and bled. Parent–child pairs have the advantage of avoiding any ethical considerations as an accurate diagnosis has a realistic chance of benefiting the patient from any therapy or preventive action which might be discovered in time to slow the progression and delay manifestation in future cases, especially unborn sibs. If as is usual, the mother accompanies the child, the occasional problems of either concealing or handling an inconsistent paternity could not arise. There would be no within-family linkage data and the between-family linkage information would require the family to be 'local' to the population under study if haplotype comparisons are to be defined. The procedure is useless for the analysis of loose linkage, but ideally suited for checking on 'candidate loci', which by definition are 'the marker' or are known to be very close from other data, including that from comparative mapping. It has the advantage of only finding linkages close enough to provide a realistically short list of candidate loci.

Both parents will always provide more information than one parent by allowing the deduction of the origin of more markers and providing a second normal haplotype. Here I only consider the purely statistical information on linkage per unit resource from fully informative families. A second parent will rarely provide more information than an additional half and the laboratory workload will be increased by a half. In practice it seems reasonable to assume that the cost of resources for twosomes compared to foursomes would be of the order of a fifth, or even a tenth, at the clinical level, while at the bench it will be a half. In addition the total available resource would be much larger since only a minority of children have an affected sib.

Analysis consists in listing all the markers from affected individuals, usually children, and those only present in the parent, and looking for a difference. If the common carrier ancestor were g generations back the expectation of any haplotype having an identical allele from the common ancestor would be:

$$u = (1-t)^{2g} \qquad \textbf{(Eq. 5.5)}$$

which when t is small (< 0.1) is very close to $(1-2t)^g$.

A very distant ancestor would be expected to have had at least one recombinant or to have been born before this mutation occurred, and have the same allele with an expectation q, the allele proportion. The former is often termed 'identical by descent' or IBD and the latter 'identical by state' or IBS, but the terms have the disadvantage of being distinguished by uncertain information: all alleles are identical by descent and recent descent merely relates to the depth of family memory or local records.

The $2n$ markers in the affected individual would have an expectation of identity of:

$$u+(1-u)q \text{ where } u = (1-t)^{2g} \qquad \textbf{(Eq. 5.6)}$$

while the 'non-transmitted' parental allele would have an expectation of q, the proportion of the most frequent marker allele on the affected chromosomes.

If we select the allele that maximises the difference, however defined, in the two sets, we may consider the chance of this being fortuitous and make allowance for this selection later. In practice the allele of interest will be the commonest in the affected set in mendelian, as opposed to multifactorial, disorders.

If a is defined as the proportion of identical alleles in the gametes conveying the mutant and b the proportion lacking it, and c and d the equivalent in the population or 'control' gametes:

$a = q+u(1-q)$ where $u = (1-t)^{2g}$
$b = 1-a$
$c = q$
$d = 1-q$ **(Eq. 5.7)**

giving a cross ratio and variance

$x = (ad)/(bc)$
$v = (1/n) (1/2a+1/2b+1/c+1/d)$ **(Eq. 5.8)**

Then if

$y = \log_e(x)$
$\chi = y/\sqrt{v}$

Where numbers are small, Haldane (1956) suggested a simple continuity correction for small numbers by adding a half to a, b, c and d to compute the cross ratio and one to compute the variance. Comparison of efficiency with the four-somes is considered after discussion on the 'orphans' below.

Population studies: orphans

Given a set of affected and unaffected individuals we can study the affected for homozygous segments, with or without the benefit of controls, since the remaining genome acts as a partial control, and compare the frequencies of alleles in the affected and control population.

Homozygosity

The expectation of a locus being homozygous is $(1-h)$ where h, the usual symbol for heterozygosity, is the expected proportion of heterozygotes.

The expectation of a series of k loci all being homozygous is proportional to

$(1-h)^k$

or since *h* is likely to vary by locus, to the product of the expected homozygosities over all *k* loci.

It is the ratio of these products which provides the power of simple homozygosity mapping in between-family analysis which is allele-specific and within family analysis which is not. In any individual there will be many segments of homozygosity, and with heterozygosity levels as low as one-half, segments defined by markers too distant to show appreciable allelic association will have an expectation of homozygosity of 1/2, 1/4, 1/8 for one, two and three markers, and for the heterozygosities of 1/5, now common with multiallelic markers, 1/5, 1/25, 1/125 etc. Allelic association will moderate this rate of decay and will be more intense in the segment ascertained for homozygosity for a rare allele, as in recessive disorders. Segments ascertained through a defined rare allele differ from other segments in not only being homozygous, but in being homozygous for the same allele.

In practice a preliminary screen for segments of homozygosity or 'bald patches' is simple and merely requires the enumeration over all individuals of each marker and the number of homozygous markers with which it is contiguous. The 'winning' loci are then examined for allele identity, a necessary feature only of segments defined by a single rare homozygote, which on average will be centrally placed. Examples are shown below. Although calculation would be simple if the necessary parameters were known this can be dispensed with in practice and retrospectively considered after examining sets of haplotypes, with and without the mutant allele, for longer segments of identical homozygosity in the former. This is not a rigorous approach in that it does not allow exact inferences since the test is decided after examining the data. This is also the standard procedure in maximum likelihood approaches to linkage analysis when its advantages usually outweigh its bias. Examples of graphical representation are shown below.

Allelic association

The haplotype including the mutant allele will have an expectation of a marker allele with proportion *q* and distance *t* cM of:

$$a = q+u(1-q) \text{ where } u = (1-t)^{2g}; \ b = 1-a$$
$$c = q; \ d = 1-q;$$
$$x = (ad)/(bc)$$
$$v = (1/n) \ (1/2a+1/2b+1/2rc+1/2rd)$$
$$y = \log_e(x); \ \chi = y/\sqrt{v} \qquad\qquad \textbf{(Eq. 5.9)}$$

The analysis follows that described for twosomes apart from the smaller variance due to two control alleles rather than one, with the option of increasing this by a factor r, the ratio of controls to affected. Data already available on controls may exceed that on the affected. Even if no data are available it may be economic to have more controls as these are usually easier to acquire. The optimal 'signal-noise' ratio will depend on the relative cost of acquiring affected and control data. The information will be approximately proportional to the reciprocal of the sum of their reciprocals. For 100 affecteds and controls this is $1/(0.01 + 0.01)$ or 50 while doubling the controls would give $1/(0.01 + 0.005)$ or 67, an increase of 34% for 50% more tests, but often less than 34% more work. However the benefits soon decline. With an infinite number of controls the information is only doubled. The analyses below assume equal numbers ($r = 1$).

The cost per unit of data collection is small, and the procedure is tolerant of diagnostic errors and difficulties, which reduce the efficiency but cannot lead to more erroneous conclusions than chance. Stratification by area and, where available, by inferences based on surname and religion etc, however vague, can but increase the power, provided classification is made in ignorance of the genotype. Woolf (1957) summarised the appropriate weighted estimation procedure.

An obvious statistical problem is that the offending allele is defined after examining the data, but this is sufficiently resolved by multiplying the apparent probability from standard statistical tests by the number of alleles, or more exactly, but less conservatively, by computing $1-(1-p)^k$ where there are k alleles. As a single allelic association is under scrutiny multiple degrees of freedom are not involved. Each allele is compared with the set of all other alleles leading to a series of 2×2 tests, one for each allele. In mendelian disorders, as opposed to multifactorial disorders when resistance alleles are important, only the commonest allele in the affected set needs consideration. The null hypothesis relates to a single allele in a set of alleles being over-represented due to being present at the time of the mutation or at the time when the common ancestor migrated. The generalised distortion for which the conventional test involving comparing expectations with observations over k^2 calls for $(k-1)^2$ degrees of freedom. In this context this is inappropriate and inefficient as only one of the many degrees of freedom is relevant.

The advantage of orphans over parent–child pairs is simplicity in ascertainment and the opportunity for routine procedures at autopsy, including retrospective studies on old sections. Given rare alleles very near the offending locus it is more powerful than the twosome approach, but lacks the ability to use several markers due to the impossibility of haplotyping. This is a very major shortcoming. Data on allelic association are available from foursomes when there are two control alleles, and the power for single markers is as for orphans with an equal number of controls, with the additional advantage of haplotyping.

An example of the success of allelic association in a recessive disorder is the localisation of congenital adrenal hyperplasia to the HLA complex and the very close linkage inferred from a common haplotype in Ireland, Norway and Iceland, whose major ancestry was from Norwegians and Oceanic Celts (Arnason *et al.*, 1977).

Optimal strategy

Figure 5.2 shows the relative power of foursomes, twosomes and orphans for 100 children. The greater power of the foursomes is for recombination fractions exceeding 0.1 for a 20-generation and 0.2 for a ten-generation coancestry.

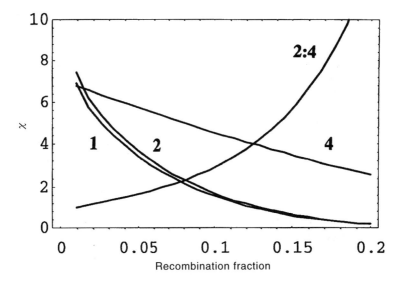

Figure 5.2 Values of χ for 100 affected children analysed as sib pairs, twosomes and orphans, marked 4, 2 and 1, and the ratio of χ for twosomes and foursomes (2:4), based on a common ancestor five generations back.

If, as seems reasonable, the resources necessary to acquire one foursome is equal to that for five twosomes, or ten orphans, and the common ancestor is ten generations back, then for linkages within 10 cM, both twosomes and orphans are more efficient; for five generations, however, the foursomes only become more efficient below about 20 cM. The number of generations is probably rather generous, especially for Nordic countries and parts of India where there has been isolation through poor transport in the past, compounded by physical and cultural barriers. This may reduce the average, or more exactly the harmonic mean, of the distance between randomly affected homozygotes in the same country to their carrier ancestor in a recessive condition to as little as five, even though the common ancestor to all homozygotes with the same mutation may be hundreds of generations distant. Since the meioses are not bounded by the number of immediate parents, the typing of neighbouring loci after selection by a suspiciously close neighbour is efficient, although some increase in significance will accompany false clues.

The twosomes, while marginally inferior to the orphans on single-marker analysis, have the potential of far greater power by providing haplotypes. In general it would seem that they provide the optimal strategy for the mapping of recessive disorders with the resources now available. Preferentially untransmitted alleles in sibships ascertained for multiple affected cases provide similar evidence. Data from orphans lack the advantage of haplotyping but can always be added through weighted estimates of the logarithm of the cross ratio. In multifactorial disorders twosomes and foursomes will usually be less efficient than orphan studies due to other loci affecting the phenotype.

The prospect of a major study conducted at a high clinical level on a recessive disorder achieving the thresholds of credulity sometimes demanded by referees and editors, or recommended as guidelines in other conditions in which larger numbers would be needed, is limited. Indeed some caution is necessary in accepting any data which passes some tests of mathematical credulity on studies of realistic size in rare conditions requiring diagnostic expertise. Publication is needed with guarantees of access to all raw data. Information supporting both linkage and exclusion can then be assembled and the future built on the secure foundation of the raw data of the past, rather than on the inferences and reports of inferences that continue to be accepted for publication unsupported by raw data, and usually refereed without recourse to it. The simultaneous publication of raw data in machine-accessible formats is standard practice with DNA and protein sequences in leading journals. Fortunately the several formats for programs analysing linkage are so similar and so obvious, in that there is nothing to specify except identity, parentage, phenotype and allele, that lack of an established format is no barrier to immediate action.

Display of results using haplotypes

Juvenile Batten's disease, which has a manifestation similar to Tay–Sachs' disease, is one of the commoner forms of blindness in early childhood but neither disorder is usually considered as such since the progressive blindness is dominated by dementia. I am grateful to Dr Mitchison and her colleagues in Britain and Finland (Mitchison *et al.*, 1993: 1995) for allowing me to use these data to explore graphical approaches to analysis. The locus was first located to chromosome 16 by within-family studies using a protein marker and then defined to a short segment by DNA studies. It was finally pinpointed by between-family studies; these showed the survival of an unbroken segment of chromosome over at least 2500 years representing at least 200 meioses connecting the common ancestor in distant branches of the family.

Haplotype analysis is robust in relation to locus heterogeneity, as well as having the ability to detect and locate multiple loci without bias in the presence of very close linkage. Although unaffected sibs provide little information if a single locus is present, they provide some additional information for haplotyping and may provide crucial information on heterogeneity if they share a segment that cosegregates with the disease in other families. They also add to the information available on the order and position of the marker loci. The analysis of what may be termed 'untidy families', since they do not conform to a standard form and number, is well covered by established linkage procedures which were used with success in these families. The detection of linkage and the approximate positioning of the loci was entirely due to the LINKAGE program with selection of the most likely position from the EXCLUDE program, which provides a simple graphical display from pairwise data using both the positive information of linkage and the negative information against linkage (Edwards, 1987).

The data can be analysed further either by counting the same and different alleles at each of the loci, here abbreviated to A, B, C . . ., which form the haplotypes of the children, when the expected proportion of a different allele will be $2 t (1 - t)$ where t is the recombination fraction. For close linkage when $(1 - t)$ is effectively 1, it will be half the proportion of differences. None were observed. This, if analysed in terms of lods, is basically the original formulation of Morton (Morton, 1955) in which affected-only sibs can be reduced to his z_1 score.

The data presented here have now been extended both in number of families and number of loci, but this early subset suffices to demonstrate the power of simple haplotype displays. As the marker alleles are all codominant no information is required external to the family analysed. The alleles are defined by number. Only one locus has more than nine loci and this, the tenth, is termed 'A'. The Batten locus is charted as '*' and the normal or wild type as '+'. In comparisons the allele is shown if identical: if different a # is shown, if no information a '.'. The loci are designated alphabetically as A, B, C.

Table 5.1 shows parental haplotypes followed by their children, the set allowing the individual haplotypes to be deduced. To the right is an abstract showing the two offending haplotypes.

49

In the first family there is no evidence of identity of alleles at defined loci around the assumed location of the locus with the offending allele which is highly suggestive of an independent mutation at either the same, or even another, locus. The second family shows complete identity to the left, but at least one recombinant must have occurred in the lines of descent between the affected locus '*'. The parents appear to have conveyed the same mutation with conservation to the left but a recombinant has occurred somewhere between the disease locus and locus P. In the third family the '3162' segment must be assumed to be common in

Table 5.1. Haplotypes of parents and children

A				Family	ABCDEFGHIJKL*MNOP*R	ABCDEFGHIJKL*MNOP*R	ABCDEFGHIJKL*MNOP*R
Hc	1	M	0	103	316.63721644*...6..	325.65211464+...6..	
Hc	2	F	0	103	316.73712465*...4..	316.45512326+...2..	
H*	3	.	1	103	316.63721644*...6..	316.73712465*...4..	316.#37#####*...#..
H*	4	.	1	103	316.6.721644*...6..	316.4.512465*...4..	316.#.######*...#..
B							
Hc	1	M	0	131	2.6272112765*...2..	3.6171422466+...3..	
Hc	2	F	0	131	3.6272112765*...3..	2.2172122335+...5..	
H*	3	M	1	131	..6272112765*...2..	..6272112765*...3..	..6272112765*...#..
H*	4	F	1	131	2.6272112765*...2..	2.6272112765*...3..	2.6272112765*...#..
C							
Hc	1	M	0	14	316245212765*.1.521	316172721336+.1.411	
Hc	2	F	0	14	316273713564*.1.511	126172323337+.1.221	
H*	3	M	1	14	316245212765*.1.5.1	316273713564*.1.5.1	3162###1##6#*.1.5.1
H*	4	M	1	14	316245212765*.1.5.1	316273713564*.1.5.1	3162###1##6#*.1.5.1
H.	5	F	1	14	316.727.13.6..1.411	316.737.35.4..1.511	
D							
Hc	1	M	0	22	326275522565*.2.812	126175522656+.2.A11	
Hc	2	F	0	22	315175522465*.2.811	111145532676+.2.411	
H*	3	F	1	22	326275522565*.2.812	315175522465*.2.811	3###75522#65*.2.81#
H*	4	F	1	22	326275522565*.2.812	315175522465*.2.811	3###75522#65*.2.81#
H.	5	M	1	22	12617552265...2.A1.	11114553246...2.81.	
E							
Hc	1	M	0	25	116272512363*.1.511	325134221327+.1.611	
Hc	2	F	0	25	126272512363*.1.511	325372712635+.1.511	
H*	3	M	1	25	116272512363*.1.511	126272512363*.1.511	1#6272512363*.1.511
H*	4	F	1	25	116272512363*.1.511	126272512363*.1.511	1#6272512363*.1.511
				family	ABCDEFGHIJKL*MNOP*R	ABCDEFGHIJKL*MNOP*R	ABCDEFGHIJKL*MNOP*R

```
zzz
1    A    S159     D16S159
2    B    S294     D16S294
3    C    S167     D16S319
4    D    S67      D16S67
5    E    S295     D16S295
5    F    S296     D16S296
6    G    S297h    D16S297h
7    H    S297s    D16S297s
8    I    S148     D16S148
9    J    S288     D16S288
10   K    S298     D16S298
11   L    S299     D16S199
13   *    *        Batten
14   M    S48      D16S48
15   N    S57      D16S57
16   O    S300     D16S300
17   P    S285     D16S285
18   Q    S150     D16S150
19   R    S151x    D16S151
zzz
```

Each set of four lines represents a family, with father, mother and affected children.
Where the data permit haplotyping the two parental haplotypes are displayed with the one inferred to include the mutant allele to the left, followed by their children each with the deduced paternal haplotype to the left. The numbers identify the alleles.
'*' defines the mutant allele and '.' an untested allele or one whose origin could not be deduced.
On the right the identity of the paternal and maternal haplotypes is shown as a '#' denoting the difference.
The loci are identified alphabetically with a key below. The inferred position of the offending locus and the order of the markers are from the published data.

the limited populations from which many marriages in Finland take place or its coexistence due to two recombinants among fairly close relatives. The third shows a recombinant near the disease locus to the left but probably not to the right. The fourth family is selected for display because there is clearly at least one break, but the strange identity of the alleles 3162 suggest either a double recombinant, probably at different times in a small community over the last millenium or a recent mutation at this marker. As the loci have been selected for allelic variability and this family was selected for display mutation is the more likely explanation. The use of multiallelic markers has the anomalous feature that they will

Table 5.2. Haplotypes sorted around disease locus with various identifying and documenting numbers

ABCDEFGHIJKL*MNOP*R	ABCDEFGHIJKL*						
3bCd7f5HI363*.N.5qR ...	AB6d7f5Hi363*	20	12	0	0	0	0	4
AB6d7f5Hi363*.N.5QR ...	AB6d7f5Hi363*	26	12	3	0	3	0	5
Ab6d7f5Hi363*.N.5QR ...	Ab6d7f5Hi363*	26	12	10	0	13	0	6
3B6d7f5.i363*.N.AQR ...	i363*	18	4	4	0	17	0	7
A.6dE56Hi363*.N.PqR ...	6dE56Hi363*	49	10	4	0	21	0	8
3b6d737Hi463*.N.P.. ...	3b6d737Hi463*	29	12	2	0	23	0	9
a.6d355Hij63*.N.4QR ...	6d355Hij63*	49	1	2	0	25	0	10
3B6 637hI644*...6.. ...	637hI644*	104	8	0	0	25	0	11
3b6d.3..i344*.N.7q. ...	i344*	10	4	2	0	27	0	12
abcd.5..i344*.N.7q. ...	i344*	10	4	4	0	31	0	13
3.cd.3ghi464*...7QR ...	3ghi464*	40	7	1	0	32	0	14
.B6.63gH3364*.N.4q....	63gH3364*	37	8	2	0	34	0	15
3B6d737H3564*...5.R ...	3B6d737H3564*	15	12	2	0	36	0	16
ABcD437.i355*.n.5QR ...	i355*	18	4	0	0	36	0	17
ABcdE3g.i365*...3.. ...	i365*	116	4	1	0	37	0	18
ab6D.55..465*...3Qr ...	465*	39	3	2	0	39	0	19
3BCd455..465*...5Q. ...	465*	43	3	3	0	42	0	20
aB6d35g.I465*...7Q. ...	I465*	36	4	3	0	45	0	21
...D.35.i465*...4.R ...	i465*	92	4	3	0	48	0	22
3B6.737Hi465*...4.. ...	737Hi465*	104	8	4	0	52	0	23
3B6D35gHi465*...5qR ...	3B6D35gHi465*	32	12	5	0	57	0	24
3b5d74ghi465*...5QR ...	3b5d74ghi465*	32	12	4	0	61	0	25
3b6D35g.i465*...7qR ...	i465*	61	4	4	0	65	0	26
.....55.i465*...A.R ...	i465*	91	4	4	0	69	0	27
3b6..5ghi565*....Qr ...	5ghi565*	41	7	2	0	71	0	28
.....f3.i565*...3.R ...	i565*	91	4	4	0	75	0	29
3b6.736Hi565*...6Qr ...	736Hi565*	38	8	4	0	79	0	30
aB6d757..665*...3Q. ...	665*	43	3	2	0	81	0	31
3BcD355.3765*...7Qr ...	3765*	61	4	2	0	83	0	32
3..D.3..i765*....QR ...	i765*	42	4	3	0	86	0	33
3..D.3..i765*....qR ...	i765*	42	4	4	0	90	0	34
Ab6..5gHi765*....qR ...	5gHi765*	41	7	4	0	94	0	35
...d.57.i765*...3.R ...	i765*	90	4	4	0	98	0	36
a.6d7fGHi765*...3.. ...	6d7fGHi765*	14	10	4	0	102	0	37
a....f..i765*...4.R ...	i765*	89	4	4	0	106	0	38
...d.7g.i765*...5.R ...	i765*	90	4	4	0	110	0	39
ABcD.F5.i765*...5QR ...	i765*	57	4	4	0	114	0	40
...D.fg.i765*...6.R ...	i765*	92	4	4	0	118	0	41
aB6D.F5.i765*...6QR ...	i765*	57	4	4	0	122	0	42
..6.7f7Hi765*...6.. ...	7f7Hi765*	17	8	4	0	126	0	43
3B6.7fgHi765*...6.. ...	7fgHi765*	1	8	5	0	131	0	44
3B6d7fg.i765*...AQ. ...	765*	36	4	4	0	135	0	45
3b6D6fg.i765*...pQR ...	i765*	119	4	4	0	139	0	46
3B6.735Hi765*...p... ...	735Hi765*	1	8	4	0	143	0	47
a.6d7fGHi765*...p.. ...	6d7fGHi765*	14	10	5	0	148	0	48
3b6D.5g..865*...AQr ...	865*	39	3	2	0	150	0	49

To aid identity alleles 1 and 2 are designated by the upper and lower case letters of the locus.
Only segment to left of the locus with the mutant allele is sorted, the sort sequence being from right to left is shown.
The sort ignores all loci separated by an indeterminate allele.
The second column merely displays identities of defined compact allelic series and is merely a tidier subset of the full data designed to be easier on the eye.
The numbers define various measures of similarily although the various methods of analysis were less informative than visual inspection.

usually be far more mutable than the mutant locus, although this has also been selected for mutability by its ascertainment. In consequence any attempt to estimate recombination rates for very close linkage with a single marker without the support of its neighbours is likely to be in error unless allelic association is absolute.

These families were selected to show atypical features: over the whole set the remarkable feature was the rarity of full homozygotes, that is homozygosity extending over the whole segment, the only example being the last family shown in spite of the rarity of recombination. This suggests the presence of an embryonic lethal, probably a deficiency, fairly close to this segment which was established before the mutation.

Substantial data for close linkage can be provided by the more numerous sporadic cases, since mother–child pairs will allow the haplotype to be deduced. *Table 5.2* shows the affected haplotypes from the disease locus. In the example there is a sort to the left. Due to the differences between most haplotypes within sibships the extra information from 'breaking up' the families to reveal their connecting haplotypes is obvious.

It is simple to 'move' the offending locus and count the number of neighbouring identities after a zig-zag sort around the locus. *Table 5.3* shows the result. Once the order is defined the haplotypes can be compared with each other in a triangle, where each haplotype is compared with every other, as in distance diagrams between towns: the number of identities astride the locus is shown (*Table 5.4*). The country of origin is to the right. There are clearly several distinct

Table 5.3. Automatic haplotype similarity counts for various positions of the disease locus

A*CDEFGHIJLKNOPQR	2	72	218	290		0.04	0.53
AB*DEFGHIJLKNOPQR	3	154	159	313		0.06	0.84
ABC*EFGHIJLKNOPQR	4	208	113	321		0.12	1.19
ABCD*FGHIJLKNOPQR	5	296	97	393		0.17	1.49
ABCDE*GHIJLKNOPQR	6	310	67	377		0.25	2.64
ABCDEF*HIJLKNOPQR	7	322	89	411		0.31	2.09
ABCDEFG*IJLKNOPQR	8	287	142	429		0.44	1.05
ABCDEFGH*JLKNOPQR	9	285	97	382		0.48	1.47
ABCDEFGHI*LKNOPQR	10	284	47	331		0.46	3.02
ABCDEFGHIJ*KNOPQR	11	250	76	326		0.52	2.12
ABCDEFGHIJL*NOPQR	12	249	99	348		0.55	1.91
ABCDEFGHIJLK*OPQR	13	261	96	357		0.51	1.90
ABCDEFGHIJLKN*PQR	14	228	29	257		0.50	4.93
ABCDEFGHIJLKNO*QR	15	193	93	286		0.52	0.52
ABCDEFGHIJLKNOP*R	16	184	45	229		0.47	0.33
123456789.1234567	jz	left	right	sum	tail	left	right

Table 5.4. Haplotype pairwise comparisons

```
 1   165   ..6.55622663*.A..    4\  Fil
 2    62   316177522373*25..    ..\  Ic
 3    28   116274212473*13..    ...\
 4     1   3262.3..2344*172.    ....\   Ne
 5     1   2222.5..2344*172.    ....4\  Ne
 6   121   .....52..464*.5..    .......\  US
 7    46   125.71712664*261.    .......\  UK
 8    17   1121437.2355*2511    ........\  Ge
 9    57   .26155..2365*.1..    .........\  US
10    15   3211745.7265*1611    ..........\         Ge
11    47   2.6235512236*1411    ..........\         Ne
12    19   321272511336*1521    ...........\        Ge
13    17   3162725.2336*1A11    ...........3\       Ge
14    25   116272512336*1511    ...........34\      Ge
15    25   126272512336*1511    ...........349\     Ge
16    47   1.6215612336*1A21    ...........3455\    Ge
17    28   326273712436*11..    ................\
18    31   .16263213346*142.    ................\      UK
19    39   3.22.3222446*.711    ................\      Ne
20    14   3162737135 46*1511   ................\      UK
21    44   32217721.256*1511    ...................        No
22    44   21617721.256*1211    ...................3\     No
23    36   322172613256*.911    ...................33\    Ge
24   119   .....22..356*.A..    ........................\       US
25   104   1122132.2356*.3..    ........................3\      US
26    15   1262757.2356*1311    ........................34\     UK
27    13   316263512356*1312    ........................344\    UK
28    27   1262632.1456*1511    ...........................\        Sw
29    32   2162352.1456*.711    ...........................4\       UK
30    37   3122455.1456*.51.    ...........................44\      Ne
31     6   3261352.2456*.721    ...........................333\     Ne
32    91   ...1.35.2456*.4.1    ...........................3334\    De
33    90   .....55.2456*.A.1    ...........................3344\    De
34     3   316135212456*.521    ...........................333444\
35    18   316145712456*1421    ...........................3334445\    Ne
36     3   325274222456*.511    ...........................33344444\   Ne
37    22   315175522456*2811    ...........................333444445\  Ne
38    43   325272622456*1A..    ...........................3334444455\ No
39    38   2261.55.3456*.312    ...........................33333333333\ Sw
40    26   126262211556*25.1    ...........................\
41    90   .....23..2556*.3.1   ...............................3\            De
42    24   3161727.2556*1211    ...............................34\           De
43    92   .....37.2556*...1    ...............................344\          De
44    46   316.61512556*2A1.    ...............................3444\         Ne
45    36   326273612556*.612    ...............................34445\        Ge
46    40   3262.5222556*...12   ...............................344444\       Ne
47    22   326275522556*2812    ...............................3444445\      Ne
48    31   .16155722556*151.    ...............................34444455\     US
49    29   215235231656*221.    ...................................3\         US
50    57   .12235..2656*.A..    ...................................3\          US
51    37   2162757.2656*.31.    ...................................34\         Sw
52   121   .....32..756*.9..    ...................................\          US
53   119   .....52..756*.6..    ...................................3\          US
54    45   316.63221756*1511    ...................................33\         Sw
55    23   226272421756*15..    ...................................335\        Ne
56    18   216235521756*2512    ...................................3355\       Ne
57    45   326.73721756*1221    ...................................33555\      Sw
58    20   312275721756*1511    ...................................335556\     Ne
59    61   2..1.2..2756*.4.1    ...................................3333333\        UK
60    41   3..1.3..2756*..11    ...................................33333334\       Ne
61    41   3..1.3..2756*..21    ...................................33333344\       Ne
62    91   ...1.22.2756*.6.1    ...................................3333333444\     De
63    32   3162722.2756*.A12    ...................................3333333344\     UK
64    89   ...2.72.2756*.5.1    ...................................33333333444\    Fil
65    10   3261672.2756*.211    ...................................3333333344444\  Ne
66    42   1121.15.2756*.512    ...................................33333334444444\ Ge
67    42   2161.15.2756*.611    ...................................33333334444444\ Ge
68    24   3222745.2756*1711    ...................................33333334444444\ De
69    27   3162737.2756*2222    ...................................3333333444444444\ Sw
70    89   ...2.57.2756*.3.1    ...................................3333333444444444\ Fil
71   131   2.6272112756*.2..    ...................................333333344444444444\       US
72   131   3.6272112756*.3..    ...................................333333344444444444\       NF
73    70   316172212756*.6..    ...................................3333333444444444455\      Fil
74    52   226172212756*1512    ...................................3333333444444444559\      Fil
75    53   312272212756*25..    ...................................33333334444444444455 88\   Fil
76    52   116272212756*1511    ...................................33333334444444444455 889\  Fil
77    49   316272212756*1622    ...................................33333334444444444558899\   Fil
78    49   115274212756*1512    ...................................33333334444444444455566 6\  Fil
79    40   1261.5212756*..21    ...................................3333333444444444455566 66\ Ne
80    14   316245212756*1521    ...................................3333333444444444455566 667\ UK
81    70   316273512756*.2..    ...................................3333333444444444455555 5555\ US
82    13   316274512756*1621    ...................................3333333444444444455555 5555\ UK
83   165   ..6.72712756*.6..    ...................................3333333444444444455555 5555558\ Fil
84    53   316172712756*1A..    ...................................3333333444444444455555 5555558\ Fil
85    43   326155122756*1A..    ...................................333333344444444444444 44444444\ Ge
86    26   222263422756*13.2    ...................................333333344444444444444 4444445\ Sw
87    21   326243722756*1611    ...................................333333344444444444444 44444444455\ Ne
88     6   3121355.3756*.712    ...................................3333333333333333333 333333333333333\Ne
                                123456789 123456789 123456789 123456789 123456789 123456789 123456789 123456789 123456789
                                     10        20        30        40        50        60        70        80
```

Each haplotype is compared with all others after a zig-zag sort about the disease locus.
If three or more alleles at loci next to this locus show an uninterrupted identity the number is given if it exceeds two.
The countries of origin are shown to the right. Finland is emphasised by a vertical line.
De, Denmark; Fi, Finland; Ge, Germany; Ic, Iceland; Ne, Netherlands; No, Norway; Sw, Sweden; UK, UK; US, USA.

groups of haplotypes, which probably relate to different mutations, and show the expected regional distribution.

The position of the locus was 'given', as it had already been inferred by the authors. Their first set of data showed an obvious anomaly with the rightmost two loci showing frequent identity with non-identity of its neighbour to the left. The authors had independently revised this order on the second set of data. The procedure provides a very powerful method for confirming an order and suggesting small changes. It is especially suitable for recessives in farm animals when a single mutation can be assumed or for translocations. The position of the offending locus had already been inferred when I had access to the data and is now fully substantiated.

If analysis is restricted to very close markers and enriched by further data, histograms can be drawn by country of the distribution of alleles at a pair of marker loci (*Figure 5.3*). In this case it is clear that loci with alleles 6 and 5 respectively are grossly over-represented, and a map can be drawn exploiting what is known from historical evidence over the last two thousand years, and from linguistic evidence over the preceding millenium, or even millennia. It seems that the only reasonable interpretation, from linguistic evidence, is that the Finnish ancestry was somewhere around Hungary, and the offending haplotype spread preferentially with the Finns but also travelled to the North West, presumably reaching Finland by land. The British haplotype may have come mainly in the early migrations by foot, which would place the mutation several thousand years earlier between the end of the Ice Age and the washing away of the land bridge, or with the smaller seaborne migrations or the massive Saxon invasions, following the Roman invasion and colonisation, in the fourth and fifth centuries. In any case it must be several thousand years old and the accompanying loci very close and not very mutable.

Summary

Recent advances in DNA techniques allow indefinitely large numbers of multi-allelic marker loci to be defined. This makes genome scans practical in mendelian disorders.

It also imposes new problems on methods of analysis that are dependent on assumptions of few alleles, well-spaced markers and recurrent mutations.

The efficiency of various approaches is explored and it is suggested that the optimal strategy in heterogeneous communities is the affected sib pair approach, and in others either the systematic collection of parent–child pairs or association studies on affected individuals and controls. The terms foursome, twosome and orphan are advanced.

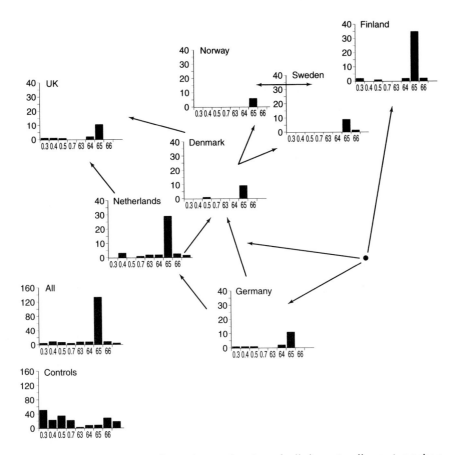

Fig. 5.3 Histograms of numbers of pairs of alleles at adjacent marker loci. The pair with alleles 5 and 6 respectively remain closely associated with the ancestral mutation, and are mapped onto the position of the countries. The arrows are inferred from the historical evidence. (Data courtesy of Mitchison *et al.*, 1993, 1995.)

The information on position when large numbers of individuals and markers are involved is approximately proportional to the product of the square root of the number of affected persons and the number of markers. Studies with less than several dozen patients and several hundred markers are uneconomic unless analysis is restricted to a few loci that are strong candidates.

The use of high lod scores or low probabilities as a criterion for publication is inappropriate as some of the limits recommended are beyond the reach of studies conducted with high clinical standards in rare disorders. There are no satisfactory procedures for accumulating information beyond access to raw data after publication.

Routine collection of DNA from parent–child pairs is an economic, ethical and practical procedure for the study of recessive disorders. There is a need to extend cell banking facilities to DNA banking.

Appendix

The estimation of linkage follows directly from the equations given in the text. If s and d are the numbers of observations in the same and different categories then since in the absence of linkage the expectations of s and d are equal; and of $s-d$ are equal to zero; and, in the presence of complete linkage, d is zero.

If $u = (s - d)/(s + d)$ then, in the absence of linkage ($t = 0.5$) the expectation of u is zero. If linkage is complete ($t = 0$) then $u = 1.0$;

For all values of t $u = (1 - 2t)^2$ so that $(1 - \sqrt{u})/2$ is an estimate of t. The figure shows the relationship and the close approximation, for close linkage, of $t = 0.27u$.

Tests for significant linkage follow directly from

$$\chi = u/\sqrt{(n)}, \text{ or } \chi^2 = (s - d)^2/n$$

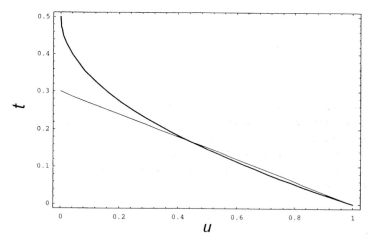

The recombination fraction, t, against u with the approximation discussed in text.

While it is conventional to express evidence of linkage in terms of z scores, these are neither needed nor appropriate when multiallelism makes counting procedures possible. While probabilities and likelihoods are formally incommensurate, the approximate lod, derived from the negative logarithm of the probability, is adequate approximation in practice. In the critical region of a probability of $1/1000$, or a likelihood ratio of $1000:1$, $z = 3.0$ and $\chi = 3.1$.

References

Arnason, A., Larsen, B., Marshall, W.H., Edwards, J.H., Macintosh, P., Olaisen, B., Teisberg, P. Very close linkage between HLA-B and Bf inferred from allelic association. (1977) *Nature* **268**: 527–528.

Edwards, J.H. Allelic association in man. In Erikson, A.W. (ed.) *Population Structure and Genetic Disorders*. pp. 239–256. (1980) Academic Press, New York.

Edwards, J.H. Exclusion mapping. (1987) *J. Med. Genet.* **24 (9)**: 539–543.

Fisher, R.A. *The Genetical Theory of Natural Selection*. (1930) Clarendon Press, Oxford,

Haldane, J.B.S. The use of linked marker genes for detecting recessive autosomal lethals in the mouse. (1955) *J. Genet.* **55**: 596.

Haldane, J.B.S. The estimation and significance of the logarithm of the ratio of frequencies. (1956) *Ann. Eugen.* **20**: 309–311.

Lander, E.S. & Kruglyak, L. Genetic dissection of complex traits: guidelines for interpreting and reporting linkage results. (1996) *Nat. Genet.* **12 (4)**: 357–358.

Lander, E.S., Schork, N.J. Genetic dissection of complex traits. (1994) *Science* **265**: 2037–2048.

Mitchison, H.M., O'Rawe, A.M., Taschner, P.E., Sandkuijl, L.A., Santavuori, P., de Vos, N., Breuning, M.H., Mole, S.E., Gardiner, R.M., Jarvela, I.E. Batten disease gene, CLN3: linkage disequilibrium mapping in the Finnish population, and analysis of European haplotypes. (1995) *Am. J. Hum. Genet.* **56**: 654–662.

Mitchison, H.M., Thompson, A.D., Mulley, J.C., Kozman, H.M., Richards, R.I., Callen, D.F., Stallings, R.L., Doggett, N.A., Attwood, J., McKay, T.R. Fine genetic mapping of the Batten disease locus (CLN3) by haplotype analysis and demonstration of allelic association with chromosome 16p microsatellite loci. (1993) *Genomics* **16**: 455–460.

Morton, N.E. Sequential tests for the detection of linkage. (1955) *Am. J. Hum. Genet.* **7**: 277–318.

Ott, J. Estimation of the recombination fraction in human pedigrees. (1974) *Am. J. Hum. Genet.* **26**: 588–589.

Risch, N. Linkage strategies for genetically complex traits. III. The effect of marker polymorphism on analysis of affected relative pairs. (1990) *Am. J. Hum. Genet.* **46 (2)**: 242–253.

Robbins, R.B. Some applications of mathematics to breeding problems. III. (1918) *Genetics* **3**: 375–389.

Sturtevant, A.H. The linear arrangement of six sex-linked factors in *Drosophila. J. Exp. Zool.* **14**: 53–59.

Tomlinson, I.P. & Bodmer, W.F. The HLA system and the analysis of multifactorial genetic disease. (1995) *Trends. Genet.* **11 (12)**: 493–498.

Woolf, B. (1957) On estimating the relation between blood group and disease. *Ann. Hum. Genet.* **21**: 397–409.

Chapter 6
Mutation-selection-equilibria, genetic models and linkage analysis

Tiemo Grimm and Bertram Müller-Myhsok

Introduction

When carrying out linkage analysis using the lod-score method, it is necessary to choose a genetic model describing the genetics of the trait that is being studied.

In the LINKAGE package the parameter file defines the genetic models for each locus in the pedigree file. Specifications can be made to sex-linked or not, to the option of mutation, to the penetrance for defining the mode of inheritance, and to the allele frequencies.

By defining all these parameters it is important to remember that under the assumption of a mutation-selection-equilibrium (Haldane, 1935), the values of these parameters are dependent upon each another. We will give examples of these dependencies and show their influence on the lod scores obtained in an analysis.

X-linked inheritance

Lethal X-linked inheritance with equal mutation rate (f = 0; μ=v)

The classical example for the mutation-selection-equilibrium is the lethal X-linked inheritance, as seen in Duchenne muscular dystrophy (DMD) (*Figure 6.1*).

GENETIC MAPPING OF DISEASE GENES
ISBN 0-12-232735-7

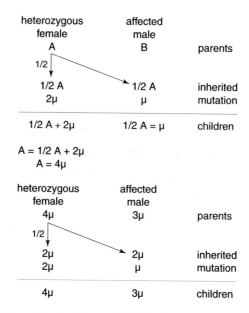

Figure 6.1 X-linked inheritance (lethal; $f = 0$; $\mu = v$)

The frequency of the heterozygotes in the parent generation ($2pq$) is A in the generation of the children it will be $1/2A + 2\mu$.
(where μ denotes male and female mutation rate).
Under the assumption of an equilibrium, the frequencies in both generations must be identical:

$$A = \tfrac{1}{2}A + 2\mu \qquad\qquad \textbf{(Eq. 6.1)}$$

and

$$A = 4\mu \qquad\qquad \textbf{(Eq. 6.2)}$$

Therefore one may write the equilibrium of *Figure 6.1* in terms of the mutation rate (μ).

This also means that the allele frequencies in females and in males (p_f, q_f, p_m, q_m denote the allele frequencies of the wildtype and disease alleles in males and females, respectively) must be different and will depend on the mutation rate:

$$2p_f\, q_f = 4\mu \qquad\qquad \textbf{(Eq. 6.3)}$$

or:

$$2q_f \approx 2\mu \qquad\qquad \textbf{(Eq. 6.4)}$$

and

$$q_m = 3\mu \qquad \textbf{(Eq. 6.5)}$$

In the parameter file of the LINKAGE package the female allele frequencies have to be entered. The following example of two families (*Figure 6.2*) demonstrates the influence of the mutation rate on the linkage analysis:

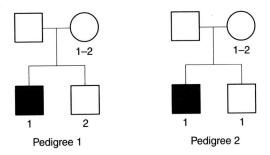

Pedigree 1   Pedigree 2

Figure 6.2 Pedigrees with X-linked inheritance

genetic model

    disease (e.g. DMD):

        – lethal recessive X-linked inheritance
        – mutation rate $\mu = 0.0001$
        therefore, following **Eq. 6.4**:
        $p_f = 0.9998$; $q_f = 0.0002$
        DNA-probe (diallelic):
        $p = 0.5$; $q = 0.5$

lod score table (*Table 6.1*)

Table 6.1. lod score table (1 = disease; 2 = DNA-probe)

Order	Pedigree	0.0	0.01	0.05	0.1	0.2	0.3	0.4
Without mutation								
1=2	1	0.30	0.29	0.26	0.21	0.13	0.06	0.02
	2	$-\infty$	-1.40	-0.72	-0.44	-0.19	-0.08	-0.02
With mutation								
1=2	1	0.18	0.17	0.15	0.12	0.07	0.03	0.01
	2	-0.30	-0.28	-0.23	-0.17	-0.09	-0.04	-0.01

Lethal X-linked inheritance with different mutation rates in both sexes ($f = 0$; $\mu < v$ or $\mu > v$)

If the mutation rates in females and males are different, a new mutation-selection-equilibrium is needed (*Figure 6.3*). In this mutation-selection-equilibrium the frequency of the heterozygotes is:

$$2p_f\,q_f = 2\mu + 2v \qquad\qquad \textbf{(Eq. 6.6)}$$

(where v denotes male mutation rate),

or:

$$q_f \approx \mu + v \qquad\qquad \textbf{(Eq. 6.7)}$$

and if:

$$k = v/\mu \qquad\qquad \textbf{(Eq. 6.8)}$$

$$q_f \approx \mu + k\mu = \mu(1 + k) \qquad\qquad \textbf{(Eq. 6.9)}$$

and

$$q_m \approx 2\mu + k\mu = \mu(2 + k) \qquad\qquad \textbf{(Eq. 6.10)}$$

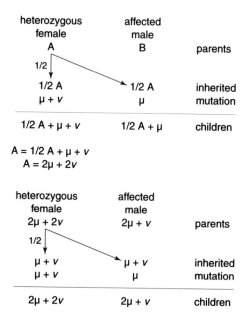

Figure 6.3 X-linked inheritance (lethal; $f = 0$; $\mu \neq v$).

In DMD Grimm *et al.* (1994) have shown that in families with point mutations the male mutation rate is about ten times higher than the female mutation rate. The genetic model will then be:

disease (e.g. DMD):

– lethal recessive X-linked inheritance
– mutation rate: $v \approx 10\mu$
$k \approx 10$; $\mu = 0.0001$; $v = 0.0010$
therefore following **Eq. 6.9**:
$q_f \approx 11\mu$; $p_f = 0.9989$; $q_f = 0.0011$.

DNA probe (diallelic):
$p = 0.5$; $q = 0.5$

Doing linkage analysis for the families of *Figure 6.2* would result in the lod score table shown (*Table 6.2*).

Table 6.2 lod score table (1 = disease; 2 = DNA-probe)

Order	Pedigree	0.0	0.01	0.05	0.1	0.2	0.3	0.4
Without mutation								
1=2	1	0.30	0.29	0.26	0.21	0.13	0.06	0.02
	2	$-\infty$	−1.40	−0.72	−0.44	−0.19	−0.08	−0.02
With mutation								
1=2	1	0.27	0.26	0.23	0.19	0.12	0.06	0.01
	2	−0.81	−0.73	−0.50	−0.34	−0.16	−0.06	−0.01

In deletion DMD families the mutation rate is about twice as high in females as it is in males (Grimm *et al.*, 1994). The genetic model will then be:

disease (e.g. DMD):

– lethal recessive X-linked inheritance
– mutation rate: $v \approx 1/2\mu$
$k \approx 1/2$; $\mu = 0.0001$; $v = 0.00005$
therefore following **Eq. 6.9**:
$q_f \approx 1.5\mu$; $p_f = 0.99985$; $q_f = 0.00015$.

DNA probe (diallelic):
$p = 0.5$; $q = 0.5$

Doing linkage analysis for the families shown in *Figure 6.2* would provide the LOD score table shown in (*Table 6.3*).

Table 6.3. lod score table (1 = disease; 2 = DNA-probe)

Order		0.0	0.01	0.05	0.1	0.2	0.3	0.4
Without mutation								
1=2	1	0.30	0.29	0.26	0.21	0.13	0.06	0.02
	2	$-\infty$	-1.40	-0.72	-0.44	-0.19	-0.08	-0.02
With mutation								
1=2	1	0.15	0.15	0.13	0.11	0.06	0.03	0.01
	2	-0.24	-0.23	-0.19	-0.14	-0.07	-0.03	-0.01

X-linked inheritance with equal mutation rate in both sexes ($f > 0$; $\mu = v$)

A mutation-selection-equilibrium for an X-linked inheritance with a reduced fertility in the affected males can be postulated ($f > 0$). A classical example is Becker muscular dystrophy BMD with a relative fertility of 0.7 (Figure 6.4). In the mutation- selection-equilibrium the frequency of the heterozygotes is:

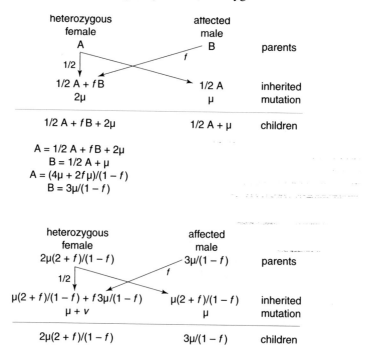

Figure 6.4 X-linked inheritance ($f > 0$; $\mu \neq v$).

$$2p_f q_f = \frac{(4\mu + 2f\mu)}{(1-f)} \qquad \textbf{(Eq. 6.11)}$$

or:

$$q_f \approx \frac{(2\mu + f\mu)}{(1-f)} \qquad \textbf{(Eq. 6.12)}$$

and:

$$q_m \approx \frac{3\mu}{(1-f)} \qquad \textbf{(Eq. 6.13)}$$

and if $f = 7/10$:

$$q_f \approx 9\mu \qquad \textbf{(Eq. 6.14)}$$

and:

$$q_m \approx 10\mu \qquad \textbf{(Eq. 6.15)}$$

The genetic model will then be:

disease (e.g. BMD):

– recessive X-linked inheritance ($f = 0.7$)
– mutation rate: $\mu = 0.0001$
therefore following **Eq. 6.14**:
$q_f \approx 9\ \mu$; $p_f = 0.9991$; $q_f = 0.0009$.

DNA-probe (diallelic):
$p = 0.5$; $q = 0.5$

Doing linkage analysis in the families of *Figure 6.2* produces the lod score table (*Table 6.4*).

Table 6.4 lod score table (1 = disease; 2 = DNA-probe)

Order	Pedigree	0.0	0.01	0.05	0.1	0.2	0.3	0.4
Without mutation								
1=2	1	0.30	0.29	0.26	0.21	0.13	0.06	0.02
	2	$-\infty$	-1.40	-0.72	-0.44	-0.19	-0.08	-0.02
With mutation								
1=2	1	0.26	0.25	0.22	0.18	0.11	0.05	0.01
	2	-0.74	-0.67	-0.47	-0.32	-0.15	-0.06	-0.01

Autosomal inheritance (Cavalli-Sforza and Bodmer, 1971).

Assuming a Hardy–Weinberg equilibrium and a diallelic model (*a* with an allele frequency of *p* and *b* with an allele frequency of *q*) results in the following conditions for the allele frequencies:

$$p + q = 1 \qquad \text{(Eq. 6.16)}$$

$$p^2 + 2pq + q^2 = 1 \qquad \text{(Eq. 6.17)}$$

The homozygotes *aa* are healthy and there is no selection against them. The fitness will then be 1. The heterozygotes *ab* have a fitness of $1-ht$ (*h* and *t* being arbitrary variables). The mutant homozygotes *bb* are affected and they have a fitness of $1-t$.

In the model for deleterious recessive inheritance $h = 0$ and $t = 1$. When $h = 1$ and $t = 1$ it is a deleterious dominant inheritance. The selection will change the Hardy–Weinberg equilibrium in the next generation (*Figure 6.5*).

genotype	fitness	frequencies before selection	frequencies after selection
AA	1	p^2	p^2
Aa	$1 - ht$	$2pq$	$2pq(1 - ht)$
aa	$1 - t$	q^2	$q^2(1 - t)$
total		1	$1 - 2pqht - q^2t$

Figure 6.5 Selection against an autosomal gene.

The allele frequencies of *a* and *b* before the selection:

$$a \to p \qquad \text{(Eq. 6.18)}$$

$$b \to q \qquad \text{(Eq. 6.19)}$$

and after selection (without mutation):

$$a \to p^1 = \frac{p^2 + pq(1 - ht)}{1 - 2htq + 2htq^2 - tq^2} = \frac{1 - q - qht + q^2ht}{1 - 2htq + 2htq^2 - tq^2} \qquad \text{(Eq. 6.20)}$$

$$b \rightarrow q^1 = \frac{pq(1-ht) + q^2(1-t)}{1 - 2htq + 2htq^2 - tq^2} = \frac{q - qht + q^2ht - q^2t}{1 - 2htq + 2htq^2 - tq^2} \qquad \textbf{(Eq. 6.21)}$$

The change of q (Δq) in one generation will be:

$$\Delta q = q^1 - q \qquad \textbf{(Eq. 6.22)}$$

$$\Delta q = \frac{-htq + (3ht - t)q^2 + (t2ht)q^3}{1 - 2htq + 2htq^2 - tq^2} \qquad \textbf{(Eq. 6.23)}$$

Δq will be lost in each generation.

If equilibrium holds, the mutation from p^1 to q^1 will occur with the mutation rate μ. If the mutation and the selection are equal:

$$\mu p^1 = -\Delta q \qquad \textbf{(Eq. 6.24)}$$

a new mutation-selection-equilibrium will exist and it is possible to write μ as a formula of q, h and t:

$$\mu p^1 + \Delta q = 0 \qquad \textbf{(Eq. 6.25)}$$

and:

$$\mu = \frac{1 - q - qht + q^2ht}{1 - 2htq + 2htq^2 - tq^2} + \frac{-htq + (3ht - t)q^2 + (t - 2ht)q^3}{1 - 2htq + 2htq^2 - tq^2} = 0 \quad \textbf{(Eq. 6.26)}$$

and:

$$\mu = q\,[ht - (3ht - t)q - (t - 2ht)q^2]\,/\,[1 - q - qht + q^2ht] \qquad \textbf{(Eq. 6.27)}$$

When h tends against 0 (e.g. in an autosomal recessive inheritance):

$$\mu \approx tq^2 \qquad \textbf{(Eq. 6.28)}$$

or when q is very small (e.g. in an autosomal dominant inheritance):

$$\mu \approx htq \qquad \textbf{(Eq. 6.29)}$$

When the mutation rate in both cases (**Eq. 6.28** and **Eq. 6.29**) is equal:

$$tq^2 \approx htq \qquad \textbf{(Eq. 6.30)}$$

or:

$$q \approx h \qquad \qquad \textbf{(Eq. 6.31)}$$

In this case the loss of mutant homozygotes equals that from heterozygotes. When $q > h$ and the influence of h on q is very small, the inheritance is autosomal recessive. When $q < h$ and the influence of h on q is important, the inheritance is autosomal dominant (*Table 6.5*).

Table 6.5. Mutation-selection-equilibrium and gene frequencies.

$t = 1; \mu = 0.0001$

q	0.010	0.0098	0.008	0.00714	0.005	0.001	0.0001
h	10^{-13}	0.0004	0.005	0.00714	0.015	0.010	1.000
	Recessive						Dominant

Autosomal dominant inheritance

Under the assumption that the allele frequency of the affected allele is very rare compared to the allele frequency of the wild type ($p \gg q$) the mutation-selection-equilibrium is as shown in *Figure 6.6*.

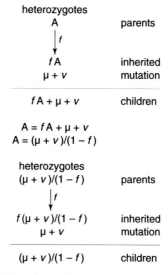

Figure 6.6 Autosomal dominant inheritance ($p \gg q$; $\mu \neq v$).

The following example of two families (*Figure 6.7*) demonstrates the influence of the mutation rate on the linkage analysis:

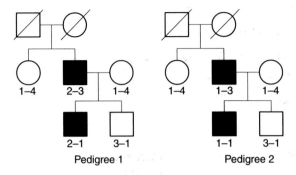

Figure 6.7 Pedigrees with autosomal dominant inheritance

genetic model:

disease:

 – autosomal dominant inheritance
 – fitness: $1-ht = f = 0.5$
 – mutation rate $\mu = 0.0001$
 therefore following **Eq. 6.29**:
 $p = 0.9998$; $q = 0.0002$;
 DNA-probe (four alleles):
 $p = 0.25$; $q = 0.25$; $r = 0.25$; $s = 0.25$
 (where p, q, r and s are the frequencies of the respective alleles)

lod score table (*Table 6.6*)

Table 6.6. lod score table (1 = disease; 2 = DNA-probe)

Order	Pedigree	0.0	0.01	0.05	0.1	0.2	0.3	0.4
Without mutation								
1=2	1	0.60	0.58	0.52	0.43	0.27	0.13	0.03
	2	−0.03	−0.02	0.01	0.03	0.04	0.03	0.01
With mutation								
1=2	1	0.48	0.46	0.41	0.34	0.21	0.10	0.03
	2	0.16	0.16	0.15	0.13	0.09	0.05	0.01

Autosomal dominant inheritance with incomplete inheritance

If in addition to the assumption of an autosomal dominant inheritance the possibility of incomplete penetrance (P) is taken into account the mutation rate is:

$$\mu = qP(1-f) \tag{Eq. 6.32}$$

Doing linkage analysis in the families of *Figure 6.7* with the above model results in the lod score table.

genetic model:

disease:

    –autosomal dominant inheritance
    –incomplete penetrance: $P = 0.8$
    –fitness: $1-ht = f = 0.5$
    mutation rate $\mu = 0.0001$
    therefore following **Eq. 6.29**:
    $p = 0.99975; q = 0.00025$
    DNA-probe (four alleles):
    $p = 0.25; q = 0.25; r = 0.25; s = 0.25$

lod score table (*Table 6.7*)

Table 6.7. lod score table (1 = disease; 2 = DNA-probe).

Order	Pedigree	0.0	0.01	0.05	0.1	0.2	0.3	0.4
Without mutation								
1=2	1	0.44	0.43	0.37	0.31	0.19	0.09	0.02
	2	0.03	0.03	0.04	0.04	0.03	0.02	0.01
With mutation								
1=2	1	0.37	0.36	0.31	0.25	0.15	0.07	0.02
	2	0.12	0.11	0.10	0.09	0.06	0.03	0.01

Autosomal recessive inheritance

In the past new mutations in autosomal recessive inheritance would tend to be neglected, but molecular analysis has demonstrated that mutations also occur in autosomal recessive disorders (Wirth *et al*, 1995).

    The following example of two families (*Figure 6.7.9*) demonstrates the influence of the mutation rate on the linkage analysis:

genetic model:

disease:

– autosomal recessive inheritance
– fitness: $1-t = f = 0$
– mutation rate $\mu = 0.0001$
therefore following **Eq. 6.28**:
$p = 0.99$; $q = 0.01$
DNA-probe (four alleles):
$p = 0.25$; $q = 0.25$; $r = 0.25$; $s = 0.25$

lod score table (*Table 6.8*)

Table 6.8. lod score table (1 = disease; 2 = DNA-probe).

Order	Pedigree	0.0	0.01	0.05	0.1	0.2	0.3	0.4
Without mutation								
1=2	1	0.25	0.24	0.21	0.17	0.10	0.05	0.01
	2	$-\infty$	-1.28	-0.62	-0.36	-0.14	-0.05	-0.01
With mutation								
1=2	1	0.24	0.23	0.20	0.16	0.10	0.04	0.01
	2	-1.46	-1.07	-0.57	-0.34	-0.14	-0.05	-0.01

Conclusions

All these examples demonstrate that the correct genetic model has an important influence on the linkage data. Therefore before carrying out linkage analysis a correct estimation of the genetic model is needed, including the estimation of the mutation rates, these results may have a profound impact on the LOD scores.

References

Cavalli-Sforza, L.L., Bodmer, W.F. *The Genetics Of Human Populations.* (1971) San Francisco, W.H. Freeman & Co.
Grimm, T., Meng, G., Liechti Gallati, S., Bettecken, T., Muller, C.R., Muller, B. On the origin of deletions and point mutations in Duchenne muscular dystrophy: most deletions arise in oogenesis and most point mutations result from events in spermatogenesis. (1994) *J. Med. Genet.* **31** (3): 183–186.

Haldane, J.B.S. The rate of spontaneous mutations of a human gene. (1935) *J. Genet.* **31:** 317–326.

Wirth, B., Hahnen, E., Morgan, K., Di Donato, C.J., Dadze, A., Rudnik Schöneborn, S., Simard, L.R., Zerres, K., Burghes, A.H. Allelic association and deletions in autosomal recessive proximal spinal muscular atrophy: association of marker genotype with disease severity and candidate cDNAs. (1995) *Hum. Mol. Genet.* **4 (8):** 1273–1284.

Chapter 7
Simple likelihood and probability calculations for linkage analysis

Cedric A. B. Smith and David A. Stephens

Summary

Linkage analysis using all available pedigree data can be extremely complicated. We point out that by suitably restricting the data to the most informative nuclear families, it may be possible to make a simple analysis for linkage between two loci without too much loss of information. The method can be readily extended to cases of heterogeneity where there may be more than one cause of the disease or other apparently inherited character under consideration.

The history of human linkage mapping

Although linkage analysis in experimental animals has a long history, its extension to human data has been slow for several reasons: there were very few identified loci of common polymorphisms until recently; human data were complicated by the different sizes of sibships and by matings that could be far from optimal from the geneticists point of view; and it is difficult to collect adequate data. Bell and Haldane (1937) found the first human linkage, between haemophilia and colour blindness (loci on the X chromosome). The first autosomal linkage, between the Lutheran blood group and secretor character was only found in 1954 by Mohr, and this was closely followed by Morton's 1956 discovery of heterogeneous linkage

GENETIC MAPPING OF DISEASE GENES
ISBN 0-12-232735-7

between elliptocytosis and rhesus blood group. The discovery of new blood groups, mainly in the post-war period, the growth of human biochemical genetics around 1960, and the later rapid growth of molecular genetics have resulted in the provision of many more loci. Also, medical investigation and the improvement in chromosome recognition have found that many inherited diseases and abnormalities are due to genes at specific loci or chromosomal aberrations.

Theoretical methods of tackling the problem of detecting human linkage began to take shape around 1935, with Fisher's papers (1935a,b) introducing u-functions and Penrose's sib-pair method (Penrose, 1935). Much work was carried out by Fisher and his colleagues and is explained in Bailey (1961). But the theory of detection and estimation of linkage was revolutionized by Morton's lod score (Morton, 1955). With the advent of computers, their use in analysing linkage data was initiated by Renwick and Schulze (1961a), and several computer programs have since been put forward. A difference between female and male recombination fractions was noted by Renwick and Schulze (1961b), but was largely ignored until recently.

The status of linkage analysis

The complete and careful analysis of linkage data can give rise to very difficult statistical problems and may involve considerable effort and time. In the days when linkage data were scarce and not easily obtainable, such a detailed analysis was often undertaken. With the present advances in human genetics and the increase in the numbers of geneticists undertaking linkage investigations, such an approach often seems hardly practical. The purpose of this paper is to show that a much simpler approach is possible provided that a less accurate conclusion can be accepted. This is not to be taken to compete with a detailed analysis, for in suitable cases such a detailed analysis may also be profitably undertaken.

Another important practical point is that many geneticists will be primarily trained in medicine and biology. Their mathematical and statistical knowledge may be limited, for nobody can be an expert in every subject. Therefore an approach using comparatively simple mathematical and statistical tools may be helpful.

Outline of the method

The approach we adopt is to confine ourselves to nuclear families, that is, families consisting of two parents and some children, with possible information also about phase using grandparents. We restrict ourselves to families in which the genotypes of one or (preferably) both of the parents are known, except possibly for the phase. (Unfortunately, this situation is somewhat idealized as genotype data for crucial individuals are often not available.)

This eliminates a number of complications and difficulties involved in the complete analysis of the data. We no longer need concern ourselves with allele frequencies, which may in some cases be unknown or doubtful. The use of data on cousins without data about the connection between them via their parents leads to complicated algebraic expressions taking into account all possible intermediate genotypes. A restriction to parents and children avoids this, though it may lead to conclusions being not quite so accurate, and conceivably to slight bias.

The data will be presented in pedigrees. Each pedigree may contain one or more nuclear families. These families can overlap, in that a parent in one family may be a child in another. The word 'family' means 'nuclear family' in the following description.

We illustrate the method with data on the linkage between tuberous sclerosis (TS) and the locus S85 on chromosome 16. For a detailed analysis of these data, see Smith and Stephens (1996). For a comprehensive account of methods of linkage analysis, see Ott (1991).

Description of the method

We assume that we have to deal with two loci. One is a locus for a 'main character', such as an inherited disease. The second locus is that of a test character, such as a blood group or biochemical or molecular marker which was tested for during the examination of the families. We make some idealizations in our assumptions such as that the locus of the main character may be assumed to be chosen at random from all the loci on the genome (except for X-linked characters, when it will be assumed that the locus is randomly placed on the X chromosome). We neglect, as usually relatively small in effect, disturbances caused by misclassification, mutation, reduced fertility and similar causes, assuming that the Mendelian laws and their extension to linkage hold.

We denote the alleles at the main character locus as $G_1, G_2, \ldots$, and those at the test character by $T_1, T_2, \ldots$ The most frequent case considered will be that of a dominant rare character with $G_1 G_1$ normal and $G_1 G_2$ affected, and $G_2 G_2$ not observed. In any case, we assume that all loci are effectively codominant, for example $T_1 T_1$, $T_1 T_2$, and $T_2 T_2$ are all different in effect. A rare dominant can be considered as effectively codominant. In each family at each locus each child gets one allele of the main character from its mother, one allele from its father, and similarly for the test character. If the alleles that the child inherits from its mother came from the same grandparent (maternal grandmother or grandfather) that child is a nonrecombinant due to a nonrecombination in the mother. Otherwise that child is a recombinant. The recombination fraction is just the proportion of recombinant children. It follows that in order to be informative about linkage the mother has to be a double heterozygote. For example, if the child has genotype $G_1G_3T_1T_4$, the mother is $G_1G_2T_1T_2$, the mother's mother $G_1G_4T_2T_3$, we have enough information to deduce that the child got the alleles G_1 and T_1 from his or her mother, and that the G_1 allele came from his or her mother's mother and the T_1 did not, so the child is a recombinant.

In order to make such a deduction, which we may call the 'ideal case', the genotype of one of the parent's parents examined must differ from that of the parent at the main locus, and the same must be true at the test locus. In the example above the mother's G_1G_2 differs from her mother's G_1G_4, and her T_1T_2 differs from her mother's T_2T_3, so that the mother must have phase G_1T_2/G_2T_1.

Frequently, the case will be less than ideal, when the genotypes of the parents and children can be determined, the phase of a parent cannot be determined, either because the grandparents have not been examined or for other reasons, in which case it will be assumed that the two possible phases of the parent have initially equal probability.

In order to keep the calculations simple, children who have in either character the same genotype as both parents (e.g. if child and both parents all have genotype G_1G_2) must be excluded. Unfortunately this implies that the methods described here will not readily apply without modification to recessive characters, which will need further consideration. The result of weeding out such families is that we can identify which recombinations occur in the mother and which in the father, and therefore deal separately with linkage in the two sexes.

Beginning the analysis

The first step in our analysis is to count how many certain recombinants and non-recombinants can be found in the data, and therefore make a preliminary estimate of the recombination fraction.

When we have a parent of known phase, such as G_1T_1/G_2T_2, in most families we can decide for each child whether it is a recombinant or nonrecombinant.

We can often add to these numbers using other families. Suppose we have a parent $G_1G_2T_1T_2$ of unknown phase, but passing on to five children the combinations of alleles G_1T_1, G_2T_2, G_1T_1, G_1T_2 and G_2T_1. If the parent is in fact G_1T_1/G_2T_2, the first three children are nonrecombinants, the last two recombinants, whereas if the parent is G_1T_2/G_2T_1, the first three children are recombinants, with the last two nonrecombinants. Therefore the family contains at least two recombinants and at least two nonrecombinants, and one possibility of either a recombinant or a nonrecombinant. So we can summarize the family as two recombinants, two nonrecombinants and one with equal chance of being either, even though we cannot specify which of the children are recombinants, which are nonrecombinants, and which are either (before the children's genotypes are taken into account).

A division of the number of definite recombinants by the total number of recombinants and nonrecombinants gives a first estimate of the recombination fraction. If this is considerably different from 0.5, there is a prima facie case for linkage; for example we might carry out a significance test to see whether the proportion of recombinants 'differs significantly' from 0.5. Otherwise we might discard for the time being the remaining pairs of loci, on the grounds that there is no good reason to think that they are linked, though the question could be reopened when more data are available (this will considerably reduce the number of pairs of loci to be tested).

We illustrate this with data on the linkage between TS (main character) and locus S85 of chromosome 16 (test characters) taken from Smith and Stephens (1996). *Table 7.1* is a summary of the data, omitting uninformative families. Each row corresponds to a pedigree. In the first row the number 2 in the first column indicates two nonrecombinants, the 0 in the second column zero recombinants, and the 2 in the third column, under the heading ϕ_2, means that there are essentially two ϕ_2 families in the pedigree. This will be defined more precisely later, but essentially a ϕ_2 family is one in which the mother has unknown phase and there are two children who are equally likely to be both nonrecombinants or both recombinants. The last five columns show similar information about fathers.

The total of the first and second columns are 22 and 5, respectively. This means that as regards linkage of TS with S85 in the mothers there are 22 certain nonrecombinations and five certain recombinations, making a first estimate for the female recombination fraction of $5/(22 + 5) = 0.185$. This is good evidence in favour of linkage. Similarly, the estimate of $7/(11 + 7) = 0.389$ for males is also in the right direction, but less indicative of linkage when considered on its own. For linkage with locus ASS (for arginosuccinate synthetase) on chromosome 9, we have the estimates $7/(17 + 7) = 0.29$ for females, and $7/(12 + 7) = 0.37$ for males, again suggesting linkage.

Likelihood calculations

Analysis of linkage data is almost universally pursued by the use of 'likelihood'. It is not therefore entirely out of place to digress at this point to explain 'likelihood' for those geneticists who feel that statistics is a necessary evil.

77

Table 7.1. Pedigree data for the S85 marker on chromosome 16, omitting uninformative families

Females					Males				
F	f	ϕ_2	ϕ_3	ϕ_4	M	m	μ_2	μ_3	μ_4
2	0	2	0	0	1	1	0	0	0
1	1	0	0	0	0	0	0	0	0
0	0	1	0	0	0	0	0	0	0
0	0	0	0	0	0	0	1	0	0
1	2	0	0	0	0	0	0	0	0
0	0	0	0	0	0	0	1	0	0
4	0	0	1	0	2	0	0	0	0
4	0	0	0	0	3	2	0	0	0
0	0	0	0	0	0	0	0	1	0
2	0	0	0	0	0	0	0	0	0
0	0	0	0	0	0	0	1	0	0
0	1	0	0	0	1	1	0	0	0
1	1	0	0	0	0	0	0	0	0
0	0	0	0	0	1	1	0	0	0
0	0	0	0	0	1	1	0	0	0
0	0	1	0	0	0	0	0	0	0
0	0	1	0	0	0	0	0	0	0
0	0	0	0	0	0	0	1	0	1
5	0	0	0	0	0	0	0	0	0
1	0	0	0	0	0	0	0	0	0
0	0	0	0	0	0	0	1	0	0
0	0	1	0	0	1	1	0	0	0
1	0	0	0	0	1	0	0	0	0
0	0	0	0	0	0	1	0	0	0

In genetics, the 'probability' of a random event can be interpreted as the proportion of cases in which such an event will or would happen in the long run. For example, if a coin is repeatedly tossed, it falls heads in nearly half the trials, so the probability of heads is regarded as 0.5. In practice, most probabilities are conditional probabilities: *If* we toss a coin, *then* the probability of getting heads is 0.5. Similarly, *if* we have an $A_1B \times A_2B$ blood group mating, *then* the probability of a child being A_1B is 0.25. This is often referred to as a 'frequency probability' because there are other types of probability that refer to degrees of belief, which need not concern us here. Such a probability usually refers to an event that has not yet been observed. Once a child from an $A_1B \times A_2B$ mating has been tested, the probability that he or she is A_1B is either 1 or 0.

In ordinary language, the words 'probability' and 'likelihood' are synonymous. But, in statistics, Fisher (1934) invented a concept that he called 'likelihood', and the name is now universally used and has a different interpretation from probability. 'Probability' refers to an event yet to be observed, while 'likelihood' refers to a hypothesis or assumption depending on events that have been observed. To explain this, we introduce the notation f for the female recombination fraction (in mothers), and $F = 1-f$ for the nonrecombination fraction, and similarly m and $M = 1 - m$ for the males (in fathers). Suppose that we have a mating in which the mother is doubly heterozygous with known phase so one can say whether each child is a recombinant or nonrecombinant. The recombination fraction f is not yet known. Before the children are tested we can say that the probability that the first child is a recombinant is f, the probability that the first two children are recombinants is f^2, the probability that the first two children are recombinant and nonrecombinant is fF, and so on. Then, should we observe that the first child is a recombinant, we could say that 'the likelihood of the value f of the recombination fraction is f itself, for $0 \leq f \leq 0.5$'. If the first child is a recombinant and the second a nonrecombinant, then 'the likelihood of the value of the recombination fraction is fF for $0 \leq f \leq 0.5$', and so on. In other words, the likelihood is what the probability would have been of the occurrence of the results which were later observed, if this probability had been calculated before the observations were made. The value of the likelihood depends on the assumed value of f (so the likelihood is a function of f) and can be plotted as a graph against f. Fisher (1932), in a pioneering investigation, found out two things. First, the likelihood function or graph essentially contains within itself all the relevant information provided by the observations (provided that the assumptions on which it is based hold – here, the probabilities of children under linkage). Second, the value of f (or other parameter), denoted $\hat{f}$, at which the likelihood attains its maximum value (the peak of the graph), is in general a good estimate of the actual value of f. However, this definition shows that the likelihood of a particular value of the recombination fraction f is not the same as the probability of f, a point to which we return later.

How do we calculate the likelihood for a mating of unknown phase, when the mother may, for example, have three children who are all either recombinants or nonrecombinants? We will refer to this as the ϕ_3 situation. We know that if the mating had one phase such that they were all recombinants the probability of this occurring would be f^3, while in the opposite phase, with three nonrecombinants the probability would be F^3. Since the two phases are initially equally likely, the final probability is $(f^3 + F^3)/2$. This is the contribution of this family to the likelihood of f. In accordance with the rule of multiplication of independent probabilities, the likelihood of the complete data is obtained by multiplying together the likelihoods for the separate families. For convenience we introduce the notation $\phi_n = f^n + F^n$ (and similarly $\mu_n = m^n + M^n$ for males), so that the likelihood of the family considered above, which we call 'family A', is $\phi_3/2$, hence the term ϕ_3 family. Consider a second 'family B' in which the mother is of unknown phase and in which there are altogether four children, so that with equal probability any one of them is a recombinant and three are nonrecombinants or vice versa. Therefore, whatever the phase of the mother, there is certainly at least one recombinant and one nonrecombinant, and the remaining two are either both recombinants or both nonrecombinants with equal probability. This family can be treated as if it was a set of three families, one with one recombinant, one with one nonrecombinant, and one ϕ_2, and therefore has likelihood $Ff\phi_2$. If we have a pedigree that contained both families A and B, its likelihood would be the product of these, so it would count as if it was one nonrecombinant (likelihood F), one recombinant (likelihood f), one ϕ_2 family and one ϕ_3 family, with likelihood $Ff\phi_2\phi_3$. Each row of *Table 7.1* gives the composition of one pedigree, represented in this way. For example, the first row represents a pedigree with 2 non-recombinants and 2 ϕ_2 children, and hence with likelihood (as regards the mother) $F^2\phi_2^2$.

An examination of *Table 7.1* shows that in the whole data we have for females, 22 definite nonrecombinants, five definite recombinants, six ϕ_2 cases and one ϕ_3. The likelihood for the whole data is accordingly:

$$F^{22}f^5\left\{\frac{\phi_2}{2}\right\}^6\left\{\frac{\phi_3}{2}\right\} \qquad\qquad \textbf{(Eq. 7.1)}$$

We can calculate its value for various values of f. Therefore $f = 0.2$ has likelihood 9.484×10^{-10}, and $f = 0.4$ has likelihood 5.829×10^{-12}. However, likelihood theory shows that it is only the ratios of likelihoods that matter, not their absolute values, so that we are at liberty to choose to multiply all likelihoods by any constant we choose. For linkage purposes, it is most convenient to 'standardize' the likelihood so that it takes the value 1 for absence of linkage, when $f = 0.5$. We then replace f and F in **Eq. 7.1** by $2f$ and $2F$ respectively, so that **Eq. 7.1** becomes:

$$\{2F\}^{22}\{2f\}^5\left\{\frac{4\phi_2}{2}\right\}^6\left\{\frac{8\phi_3}{2}\right\} \qquad\qquad \textbf{(Eq. 7.2)}$$

This takes the value 4170 when $f = 0.2$ and 24.7 when $f = 0.4$. That is, the likelihood of a recombination value 0.2 is 4170 times the likelihood 1 of no linkage. While we have cautioned that likelihood is not the same as probability, this does strongly reinforce the idea that TS and S85 are linked.

Partly because likelihoods can reach such high values that the likelihood curve is difficult to present as a graph, it is convenient to use the 'log likelihood' or the natural log of the likelihood for plotting as a function of f. We call this $L(f)$, and its graph is shown in *Figure 7.1*. Morton (1955) uses instead the common logarithm, or logarithm to the base 10, of the standardized likelihood,

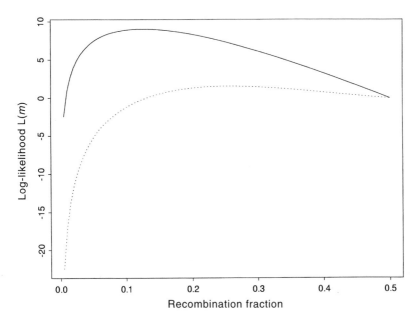

Figure 7.1 Log-likelihood surfaces for f (solid line) and m (dotted line) for the S85 locus on chromosome 16

calling this the lod (Logarithm of the odds) z. This is the natural logarithm multiplied by 0.4343. The log likelihood curve for males, $L(m)$, is also depicted in *Figure 7.1.*

The counting method

How can we find the maximum likelihood estimate of f, that is, the value of f at the peak of the likelihood (or log likelihood) curve? There is a simple technique, the counting method, suggested by Ott (1991). This depends on the definition of f as the proportion of recombinants among all children who show linkage.

Imagine, for a moment, that we have in our data just 22 recombinants and five recombinants among 27 children. The obvious estimate of f is then 5/27 = 0.185, which can be shown to be the value of f that maximizes the likelihood $L(f) = (2F)^{22}(2f)^{5}$. This suggests that we could get a good estimate if only we could count how many recombinants and nonrecombinants there are among the ϕ children. We cannot do this for certain because there is u 'hidden variation' among the ϕ families as in some the parents have one phase and in others the opposite phase, and we are unable to say for certain which are which. However, we can calculate the expected numbers of recombinants and nonrecombinants.

Consider the six ϕ_2 families in the TS and S85 data, with $6 \times 2 = 12$ children altogether. The (relative) probabilities for any such pair of children are in the ratio $F^2 : f^2$ (nonrecombinants : recombinants). If we assume for the moment that $f = 0.185$, $F = 0.815$, then this is a ratio $0.6642 : 0.0342$. To make the probabilities add to one, we must divide by the total 0.6984, giving a probability of 0.951 of two nonrecombinants in any family and 0.049 of two recombinants. Multiplying through by 12, the total number of children, gives expected numbers of 11.41 and 0.59 for nonrecombinants and recombinants respectively. Similarly, the single ϕ_3 family with three children will have relative probabilities in the ratio $F^3 : f^3$, or $0.541 : 0.006$, which after standardization represents respective probabilities 0.989 and 0.011. On multiplying by three, this gives expected values of 2.97 non-recombinants and 0.03 recombinants. Therefore the total counts now become $E_F = 22 + 11.41 + 2.97 = 36.38$ nonrecombinants and $E_f = 5 + 0.59 + 0.03 = 5.62$ recombinants. On dividing through by the total number of children, 42, this gives new estimates of $F = 36.38/42 = 0.866$, $f = 0.134$.

However, we did this calculation by assuming values of F and f that are not the true values, but only estimates, so that the new values of F and f will still only be approximations. But we expect them to be better approximations than the previous ones. We can obtain still better estimates by using $F = 0.866$, $f = 0.134$ in the calculations, which then give new estimates $F = 0.874$, $f = 0.126$. Repeating the calculation using these provisional values yields estimates $F = 0.875$, $f = 0.125$, and every further repetition of the calculation repeats these values to three decimal places, so these are our final estimates. It can be shown using calculus that these are in fact the maximum likelihood estimates, $\hat{f}$ and $\hat{F}$, corresponding to the peak of the likelihood curve.

Standard errors

An estimate of a parameter is of limited value unless we know how accurate it is. There is a standard procedure, due to Fisher, for finding the standard error of the estimate. If the downward curvature of the log likelihood curve at its peak is c, then the standard error is $\dfrac{1}{\sqrt{c}}$. This is the value provided that the log likelihood curve is approximately a parabola near the peak, as happens in the cases considered above. It is no longer applicable if the maximum is a 'boundary maximum', for example at $\hat{f} = 0$ in linkage investigations, since we cannot have a negative recombination fraction.

To find the curvature, let the log likelihood function be $L(f)$ with peak at $\hat{f}$. Calculate $L(\hat{f})$, and for (suitably chosen) sufficiently small h the log likelihoods at points at distance h on either side of $\hat{f}$ that is, $L(\hat{f} - h)$ and $L(\hat{f} + h)$. Then the curvature is:

$$c = \frac{\left[2L(\hat{f}) - L(\hat{f} - h) - L(\hat{f} + h)\right]}{h^2} \qquad \text{(Eq. 7.3)}$$

In our case the likelihood is:

$$\lambda(f) = (2F)^{22}(2f)^5(2\phi_2)^6(4\phi_3) \qquad \text{(Eq. 7.4)}$$

so that the log likelihood is:

$$L(f) = 22\ln(2F) + 5\ln(2f) + 6\ln(2\phi_2) + \ln(4\phi_3) \qquad \text{(Eq. 7.5)}$$

We have found that $\hat{f} = 0.125$. Let us take, for example, $h = 0.02$. Calculations give:

$$
\begin{aligned}
L(\hat{f}) &= L(0.125) = 9.0464 \\
L(\hat{f} - h) &= L(0.105) = 8.9703 \\
L(\hat{f} + h) &= L(0.145) = 8.9839
\end{aligned}
\qquad \text{(Eq. 7.6)}
$$

whence we find $c = 348$, s.e. $= 1/\sqrt{348} = 0.054$. It is customary to say (especially in large samples) that there is a probability around 0.95 that the true value of a parameter lies within two standard errors of the maximum-likelihood estimate. Similarly we find that the estimate $\hat{m} = 0.225$ of the male recombination fraction has a s.e. 0.083. There is controversy among statisticians about the exact interpretation of this, but virtually all agree that subject to their own interpretation it is a good approximation in large samples, and a useful guide in small samples. Therefore in the case under consideration, with about probability 0.95, the true value of f lies between $0.125 - 2 \times 0.054 = 0.017$ and $0.125 + 2 \times 0.054 = 0.233$.

In many cases there is an alternative method of finding the standard error. We have seen that (for females on chromosome 16) any ϕ_2 family has a probability 0.951 of consisting of two nonrecombinants and 0.049 of two recombinants. This is a probability distribution with variance $2 \times 0.951 \times 0.049 = 0.093$. There are six such families, giving total variance $6 \times 0.093 = 0.558$. Similarly, the single ϕ_3 family gives variance $3 \times 0.989 \times 0.011 = 0.033$. Adding these together gives a total 'hidden variance' $H = 0.558 + 0.033 = 0.591$. If E_F and E_f are the total expected counts of nonrecombinants and recombinants, we calculate:

$$j_{FF} = (E_F - H)/F^2 = (36.38 - 0.59)/0.866^2 = 48$$
$$j_{Ff} = -H/Ff = -0.591/(0.866 \times 0.134) = -5 \qquad \text{(Eq. 7.7)}$$
$$j_{ff} = (E_f - H)f^2 = (5.62 - 0.59)/0.134^2 = 280$$

and $c = j_{FF} - 2 j_{Ff} + j_{ff} = 338$ is another estimate of the curvature, giving standard error $1/\sqrt{338} = 0.054$, in close agreement with the previous value.

Heterogeneity

The correctness of these estimates is, however, thrown into doubt by the impossibility of all TS being simultaneously linked with loci on two different chromosomes. The observed data suggest that there are at least two distinct loci (namely, S 85 and ASS) for TS, one on chromosome 16, the other on chromosome 9. A full detailed discussion of this has been given by Smith and Stephens (1996). Here we adopt a simpler though less informative approach.

It is important to ask whether apparent linkage is homogeneous: that is, whether main characters such as an inherited disease are due to a single locus on one chromosome or sometimes to one locus and sometimes another locus. A clue is given by seeing that there was apparent linkage to locus S85 on chromosome 16 and ASS (locus for arginosuccinate synthetase) on chromosome 9. Smith and Stephens (1996) showed how this could be dealt with by a simple modification of the technique described here. However, a situation that is likely to occur is that when there is apparent linkage of some main character (disease) to some test character and one wants to know whether this linkage holds for all or nearly all cases of the main character or whether only some cases are linked. For example, we would have this situation if we only had the data for linkage of TS with S85 on chromosome 16, but no data concerning linkage with ASS on chromosome 9.

We now have to introduce a new parameter α, the proportion of linked cases, with $\beta = 1 - \alpha$ being the proportion of unlinked cases.

The expression for the likelihood is now more complicated. Previously, with all cases linked, we needed to take into account only the total numbers of children of varying types, such as certain recombinations, ϕ_2s etc. With possible heterogeneity, we have to calculate the likelihood for each pedigree separately and multiply them to get the likelihood for the whole data. Consider from the point of view of linkage in females the first pedigree in the chromosome 16 data. This has two definite nonrecombinant children, and two ϕ_2 children, so that the likelihood assuming linkage would be:

$$\lambda_1^{(L)}(f) = (2F)^2 (2\phi_2)^2 \qquad \text{(Eq. 7.8)}$$

In the heterogeneous case, there is an initial probability α of linkage and β of non-linkage, so the likelihood now becomes:

$$\lambda_1(f) = \alpha\lambda_1^{(L)}(f) + \beta\lambda_1^{(L)}(0.5)$$
$$= \alpha(2F)^2(2\phi_2)^2 + \beta$$

(Eq. 7.9)

where we take a standardized likelihood, which takes the value 1 when $f = 0.5$. Similarly, in the second pedigree, with one definite nonrecombinant and one definite recombinant, the likelihood is:

$$\lambda_2(f) = \alpha(2F)(2\phi_2) + \beta$$

(Eq. 7.10)

and so on. The likelihood for the whole data is then the product:

$$\lambda(f) = \lambda_1(f)\lambda_2(f)...$$

(Eq. 7.11)

In principle, we can again find maximum-likelihood estimates of f and α by a counting procedure. We make essentially two counts, one of recombinants and nonrecombinants, and one of linked and unlinked pedigrees. We have to begin with preliminary values of f and α. We could take $f = 0.125$, as previously found, so that $F = 0.875$, and α, which must lie between 0 and 1, to be 0.5, so that $\beta = 0.5$.

For each family, the likelihood splits into two terms (the first and second terms in the expression):

$$\lambda_1(f) = \alpha(2F)^2(2\phi_2)^2 + \beta$$

(Eq. 7.12)

Since we assume $f = 0.125$, $\phi_2 = 0.125^2 + 0.875^2 = 0.781$, the likelihood term for family one becomes $7.47\,\alpha + \beta$.

Using the provisional values $\alpha = \beta = 0.5$, the two terms become 3.74 and 0.5, which is interpreted as meaning that the relative probabilities that this pedigree is linked or unlinked are in the ratio 3.74 : 0.5. Dividing by the total, 4.24, we find that the probabilities of linkage and non-linkage are 0.882 and 0.118, or $P_L(1)$ and $P_N(1)$ say. This means that with this pedigree we count the expected number of linkages to be 0.882, and of non-linkages 0.118. We repeat this calculation for all pedigrees, finding a total expected number $E_\alpha = 13.06$ of linkages and $E_\beta = 9.94$ of non-linkages. Dividing by the total, 23, we find new estimates $\alpha = 0.568$ and $\beta = 0.432$.

Now we repeat the calculation to find the expected number of nonrecombinants and recombinants in the first pedigree assuming linkage by the technique already explained, namely that there are two definite nonrecombinants and two ϕ_2 cases, to be divided between nonrecombinants and recombinants in the ratio $F_2 : f_2 = 0.8752 : 0.1252$, or with probabilities 0.980 and 0.020, that is, with $4 \times 0.980 = 3.92$ nonrecombinant and $4 \times 0.02 = 0.08$ recombinant children. Thus, this gives a total count of $2 + 3.92 = 5.92$ nonrecombinant and 0.08 recombinant children, assuming linkage. But we have seen that there is a probability of linkage for this family of 0.882. Hence we modify our counts to be $0.882 \times 5.92 = 5.22$ nonrecombinants and $0.882 \times 0.08 = 0.07$ recombinants. This calculation is to be repeated for each pedigree and the estimates (on this iteration) of F and f are $E_F / (E_F + E_f) = 0.9602$ and $E_f / (E_F + E_f) = 0.0348$ respectively. Repeating the calculations using the new values of α and f as preliminary values will give more accurate estimates, and further iteration will eventually yield the maximum likelihood estimates $\hat{f} = 0$ and $\hat{\alpha} = 0.59$, suggesting that only about 60% of the pedigrees show linkage to S85.

It would be desirable to find standard errors for these estimates. The traditional method runs as follows. Let $L(f, \alpha)$ denote the log likelihood function for the whole data, and $\hat{f}$ and $\hat{\alpha}$ the maximum likelihood estimates. Choose an arbitrary small number for h,k and calculate:

$$c_{ff} = \left[2L(\hat{f}, \hat{\alpha}) - L(\hat{f} - h, \hat{\alpha}) - L(\hat{f} + h, \hat{\alpha}) \right] / h^2$$

$$c_{\alpha\alpha} = \left[2L(\hat{f}, \hat{\alpha}) - L(\hat{f}, \hat{\alpha} - k) - L(\hat{f}, \hat{\alpha} + k) \right] / k^2 \qquad \textbf{(Eq. 7.13)}$$

$$c_{f\alpha} = \left[L(\hat{f} + k, \hat{\alpha} + k) - L(\hat{f}, \hat{\alpha} + k) - L(\hat{f} + h, \hat{\alpha}) + L(\hat{f}, \hat{\alpha}) \right] / hk$$

As before, c_{ff} and $c_{\alpha\alpha}$ represent curvatures of the likelihood surface along the f and α axes, while $c_{f\alpha}$ might be called the co-curvature. We now do a matrix inversion:

$$\begin{bmatrix} v_{ff} & v_{f\alpha} \\ v_{f\alpha} & v_{\alpha\alpha} \end{bmatrix} = \begin{bmatrix} c_{ff} & c_{f\alpha} \\ c_{f\alpha} & c_{\alpha\alpha} \end{bmatrix}^{-1} \qquad \textbf{(Eq. 7.14)}$$

With only 2×2 matrices involved, the formulae for doing matrix inversions are simple. Let:

$$R = 1 / (c_{ff} c_{\alpha\alpha} - c_{f\alpha}^2) \qquad \textbf{(Eq. 7.15)}$$

Then:

$$
\begin{aligned}
v_{ff} &= Rc_{\alpha\alpha} \\
v_{\alpha\alpha} &= Rc_{ff} \\
v_{f\alpha} &= -Rc_{f\alpha}
\end{aligned}
\qquad\qquad \textbf{(Eq. 7.16)}
$$

Then v_{ff} is the error variance of the estimate, $\hat{f}$ and $s_f = \sqrt{v_{ff}}$ is the standard error. Similarly $s_\alpha = \sqrt{v_{\alpha\alpha}}$ is the standard error of the estimate, $\hat{\alpha}$ and $v_{f\alpha}$ is the error covariance.

Unfortunately, we cannot use this technique in our situation because $\hat{f} = 0$ happens to be a boundary value.

If we try a similar calculation to estimate the male recombination fraction, we run into a different problem with our data. Although the values of m and α found iteratively must eventually converge to some final stable value, the rate of convergence is painfully slow, making it impractical to estimate $\hat{m}$ and $\hat{\alpha}$ in that way.

The appropriate technique for overcoming these difficulties would seem to be to calculate the values of the log likelihood $L(f,\alpha)$ for a range of values of f from 0 to 0.5, and for α from 0 to 1, thus defining a 'likelihood surface' graphically representing the log likelihood. Such a set of values is shown in *Table 7.2*, and plotted in *Figure 7.2*.

It will be seen that the log likelihood surface has a shape very far from the customary form of a smooth single peak. Thus the usual technique using maximum likelihood estimates and standard errors to find a set of values within which the true value can be expected to lie cannot be applied here. On the contrary, for females, the likelihood surface has a sharp ridge at $f = 0$, while the male likelihood surface is rather flat, with two low peaks instead of one (*Table 7.3*, *Figure 7.3*).

Posterior probabilities of linkage

We have emphasized above that the likelihood of an event E is not the same as its probability. However, likelihoods can be used to calculate probabilities. Since the details of the argument by which this is achieved may not be generally known, it does not seem out of place to digress, presenting an example of probability calculations using Mendel (1866) and Hardy–Weinberg laws (Hardy, 1908; Weinberg, 1908).

A serologist has found that Bruno Green has blood group B. For some reason it is important for her to find out the chances of Bruno being homozygote

Table 7.2. Likelihood surface $L(f, \alpha)$ on a grid of values for the S85 marker on chromosome 16

f	0.1	0.2	0.3	0.4	α 0.5	0.6	0.7	0.8	0.9	1.0
0.00	283	3066	13653	36932	68956	91455	82061	41467	6103	0
0.02	199	2020	8821	24028	46348	65735	66914	43670	12698	52
0.04	140	1331	5679	15503	30663	45867	51423	40124	17173	765
0.06	99	877	3646	9930	20014	31248	37832	33811	18945	2775
0.08	71	580	2335	6322	12913	20868	26899	26792	18443	5390
0.10	51	384	1494	4003	8247	13700	18594	20238	16401	7509
0.12	37	255	956	2524	5221	8860	12545	14691	13572	8451
0.14	27	170	612	1586	3279	5653	8285	10303	10573	8188
0.16	20	114	392	995	2046	3565	5367	7008	7819	7093
0.18	15	77	252	623	1270	2224	3416	4638	5527	5633
0.20	11	53	163	390	784	1374	2141	2994	3755	4171
0.22	9	36	105	245	483	842	1323	1890	2463	2916
0.24	7	25	69	154	297	513	807	1169	1567	1943
0.26	5	18	45	97	182	311	487	711	971	1242
0.28	4	13	30	62	112	188	292	426	587	767
0.30	4	9	20	39	69	113	173	251	348	460
0.32	3	7	14	25	43	68	102	147	202	268
0.34	2	5	10	17	27	41	60	85	116	153
0.36	2	4	7	11	17	25	36	49	66	86
0.38	2	3	5	7	11	15	21	28	37	47
0.40	2	2	4	5	7	9	12	16	20	26
0.42	1	2	3	4	5	6	7	9	11	14
0.44	1	2	2	2	3	4	4	5	6	7
0.46	1	1	2	2	2	2	3	3	3	4
0.48	1	1	1	1	1	2	2	2	2	2
0.50	1	1	1	1	1	1	1	1	1	1

BB or heterozygote *BO*. She knows that Bruno (and his wife Olive Green) come from Ruritania. To keep calculations simple, we suppose that it is known that in Ruritania the allele frequencies of the *ABO* blood groups are 0.4:0.2:0.4 (*A:B:O*). By the Hardy–Weinberg law, a proportion $0.2^2 = 0.04$ (4%) of Ruritanians have genotype *BB*, and a proportion $2 \times 0.02 \times 0.4 = 0.16$ (16%) have genotype *BO*. Hence, on this evidence, the probability that Bruno is *BO* (conditional on finding that he has group B) is four times his probability of being *BB*. To turn these probabilities into absolute probabilities summing to 1 we have to divide by the total $0.04 + 0.16 = 0.20$, obtaining: (0.04 / 0.20 = 0.2) *BB* + (0.16 / 0.20 = 0.8) *BO*. (Alternatively, these may be expressed as $1 / (1 + 4)$ and $4 / (1 + 4)$ using the relative probabilities 1*BB* : 4*BO*.)

But now suppose that Bruno and Olive have three children and on testing all turn out to have blood group B. What are the probabilities now for Bruno?

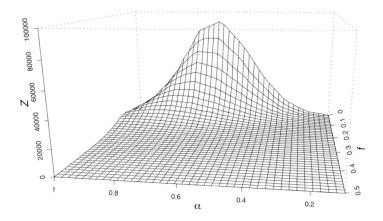

Figure 7.2 Likelihood surface for α and f for the S85 locus on chromosome 16

In Ruritania, the proportion of families consisting of a *BB* father, an *OO* mother, and three B children is (using the Hardy–Weinberg and Mendel laws): $0.22 \times 0.42 \times 1 \times 1 \times 1 = 0.0064$, while the proportion of families having a *BO* father, an *OO* mother, and three B children is: $(2 \times 0.02 \times 0.4) \times 0.42 \times 0.5 \times 0.5 \times 0.5 = 0.0032$. Comparing these figures, we see that Bruno now has probabilities in the ratio $0.0064 : 0.0032 = 2 : 1$ of being *BB* or *BO* respectively. To get the absolute probabilities, we divide by the total, giving Bruno the probabilities $2 / (2 + 1) = 2/3$ of being *BB* and $1/(2 + 1)$ of being *BO*.

We may get this result by a different route as follows. We begin by supposing that it has been found that Bruno has blood group B. As we have seen, the probabilities for Bruno are then in the ratio $0.16 \ BB : 0.64 \ BO = 1 \ BB : 4 \ BO$. These are the probabilities with which we start, accordingly called initial or prior relative probabilities.

Now suppose that the three children are tested, and found to be B. The 'likelihood' that Bruno is *BB* is defined as the probability of obtaining the observed consequences, given that Bruno is *BB* (i.e. it is $1 \times 1 \times 1 = 1$). The likelihood that Bruno is *BO* is the probability of obtaining three B children, given that Bruno is *BO* (i.e. $0.5 \times 0.5 \times 0.5 = 0.125$). Thus the ratio of likelihoods is $1 \ BB : 0.125 \ BO = 8 \ BB : 1 \ BO$. Now, multiply the prior probability ratio by the likelihood ratio: $1 \times 1 = 1 \ BB : 4 \times 0.125 = 0.5 \ BO$, or $2 \ BB : 1 \ BO$.

This is the same as the probability ratio for Bruno found previously, taking the blood groups of the children into account. These are called the final or posterior probabilities.

This is an example of what is usually called Bayes Theorem, but which is

Table 7.3. Likelihood surface $L(m, \alpha)$ on a grid of values for the S85 marker on chromosome 16

m	0.1	0.2	0.3	0.4	α 0.5	0.6	0.7	0.8	0.9	1.0
0.00	3.62	6.68	8.67	8.50	6.30	3.37	1.14	0.18	0.00	0.00
0.02	3.29	6.02	8.01	8.32	6.80	4.27	1.91	0.51	0.05	0.00
0.04	3.01	5.43	7.39	8.07	7.20	5.17	2.86	1.11	0.24	0.01
0.06	2.76	4.92	6.81	7.78	7.47	6.02	3.96	2.03	0.72	0.14
0.08	2.55	4.46	6.26	7.43	7.61	6.75	5.12	3.23	1.61	0.57
0.10	2.36	4.05	5.74	7.04	7.61	7.30	6.20	4.60	2.91	1.50
0.12	2.19	3.69	5.25	6.60	7.46	7.64	7.11	5.97	4.48	2.95
0.14	2.04	3.36	4.79	6.13	7.18	7.75	7.74	7.16	6.10	4.75
0.16	1.91	3.06	4.35	5.65	6.79	7.64	8.07	8.02	7.50	6.58
0.18	1.80	2.80	3.95	5.16	6.32	7.33	8.08	8.48	8.50	8.13
0.20	1.69	2.56	3.57	4.68	5.80	6.87	7.81	8.55	9.01	9.18
0.22	1.60	2.35	3.23	4.21	5.26	6.32	7.34	8.26	9.04	9.62
0.24	1.52	2.16	2.92	3.77	4.71	5.70	6.71	7.71	8.65	9.49
0.26	1.45	1.99	2.64	3.37	4.19	5.07	6.01	6.98	7.95	8.91
0.28	1.38	1.84	2.38	3.00	3.69	4.46	5.28	6.16	7.08	8.03
0.30	1.33	1.71	2.16	2.67	3.24	3.88	4.58	5.33	6.14	7.00
0.32	1.27	1.59	1.96	2.37	2.84	3.35	3.92	4.54	5.21	5.93
0.34	1.23	1.49	1.78	2.11	2.48	2.89	3.34	3.82	4.36	4.93
0.36	1.19	1.40	1.63	1.89	2.17	2.48	2.82	3.20	3.60	4.03
0.38	1.15	1.31	1.49	1.69	1.90	2.14	2.39	2.66	2.95	3.27
0.40	1.12	1.24	1.38	1.52	1.68	1.85	2.03	2.22	2.42	2.64
0.42	1.09	1.18	1.28	1.38	1.49	1.60	1.73	1.85	1.99	2.13
0.44	1.06	1.12	1.19	1.26	1.33	1.40	1.48	1.56	1.64	1.73
0.46	1.04	1.08	1.12	1.16	1.20	1.24	1.28	1.33	1.37	1.42
0.48	1.02	1.03	1.05	1.07	1.09	1.11	1.13	1.14	1.16	1.18
0.50	1.00	1.00	1.00	1.00	1.00	1.00	1.00	1.00	1.00	1.00

simply an example of the rule for conditional probabilities, that the (relative) posterior probabilities are obtained by multiplying the (relative) prior probabilities by the (relative) likelihoods. To get the absolute posterior probabilities, we have to multiply these by a suitable constant to make them sum to 1. To put this another way, our degree of belief in some assertion A depends on two factors; namely, how plausible it seems before the observational evidence is presented (the prior probability of A) and how probable it is that the evidence will occur if A is true (the likelihood of A).

In principle, Bayes Theorem can be used in a similar way to find the (posterior) probability of linkage, given the observations. We have already shown how to calculate the likelihood. Under our assumptions, this will be a precise value: every competent statistician asked to evaluate the likelihood should come up with the same answer.

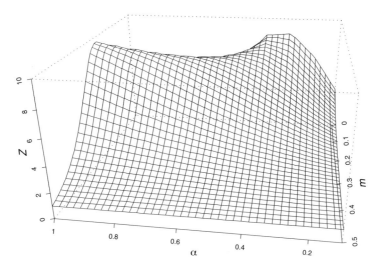

Figure 7.3 Likelihood surface for α and m for the S85 locus on chromosome 16

Prior probabilities present something of a problem in the linkage case. In essence, what we want to know is when given any particular stretch of chromosome what is the probability that the TS locus lies in this stretch before the pedigree results are taken into account? As loci seem to be distributed rather haphazardly over the genome, it is plausible to consider the TS locus as effectively a locus chosen at random from the whole set of loci. However, there is the problem that as yet we do not know precisely what is the distribution of loci.

What we can do, as a first approximation, is to consider loci as uniformly distributed over the chromosomes, where 'uniformly distributed' means that the probability of a locus lying in a given section of the chromosome is proportional to the length of that section, but there is ambiguity here. There are two distinct definitions of the length of the chromosome between two loci. The 'physical distance' is measured by the number of base pairs between the loci, while the 'map distance' is the average number of crossovers between them. Because the distribution of crossovers is irregular, with 'hot spots' where crossovers are more frequent, there is no simple relation between physical distance and map distance. Here, for convenience, as a first attempt we make the (arbitrary) assumption that the average number of crossovers between two loci is proportional to the map distance between them. (Physical distance would be better.)

It is experimentally known that crossovers never occur close together. Therefore if two loci are close together, the number of crossovers between them must be either 0 or 1. The recombination fraction between them is the probability of an odd number of crossovers, and therefore when loci are close together, the recombination fraction is equal to the map distance. When the loci are far apart, so that the map distance between them is large, observations show that the recombination fraction is close to 0.5. There have been several attempts to find a 'mapping function', that is, a universal formula relating map distance and recombination fraction. However, there seems to be no convincing reason to suppose that such a universal formula must exist. But a well-known formula relating recombination fraction θ and map distance δ that may give a sufficient approximation for our purposes and which we will use in our calculations, is that of Kosambi (1944): $2\theta = \tanh 2\delta$. Although we know that this is not true, it is not as misleading as might first appear. Even an approximate answer will give some idea of the order of magnitude of the probabilities concerned. What is more, as further information is obtained, the likelihood curve will have a taller and narrower peak, which will mean that only the distribution within this peak will become important, and any reasonable prior distribution will give approximately the same final answer. Also, the ultimate objective of the exercise is not to calculate probabilities, but to place the TS locus accurately in its place on the chromosome, and the calculation of probabilities is only a step on the road to this final answer.

Since the mothers seem to be rather more informative than the fathers, let us focus attention on the female recombination fraction on chromosome 16. Also, let us at first do the calculation on the assumption that linkage is homogeneous. Now, it has been estimated that the total (female) length of the genome is about 3400 cM, while the length of chromosome 16 is about 150 cM. So the prior probability that the TS locus is on chromosome 16 will be estimated as 150/3400 = 0.044, and the prior probability of lying on some other chromosome is 1 − 0.044 = 0.956. Furthermore, this probability of 0.044 will be supposed to be evenly distributed over the chromosome. So the prior probability that the TS locus lies in any particular 10 cM segment will be approximately 10 × 0.044 / 150 = 0.0030.

We have a slight complication in that the likelihood is calculated on the basis of the recombination fraction, but we are concerned with the position of the TS locus in cM. We convert map distance into recombination fraction using Kosambi's (1944) formula.

The likelihood will be the same whether the separation between the loci is expressed in terms of distance or recombination fraction.

The calculation is set out in *Table 7.4*. The left hand column indicates distances in cMs along the female chromosome 16 measured from the end. An entry 0.02 is to be taken as a representative of the interval between 0.00 and 0.04, an entry 0.06 represents the stretch of the chromosome between 0.04 and 0.08, and so on.

For the purposes of demonstration, we assume that the S85 locus has been found to be 10 cM from the end. The second column shows the corresponding distance from the S85 locus. This is converted using Kosambi's formula, and the likelihood calculated using **Eq. 7.2**. The likelihood is then shown in the third column. Because **Eq. 7.2** gives the likelihood standardized to make likelihood 1 in the case of no linkage, the entry opposite 'no linkage' is 1. The fourth column shows the prior probability. Since the total length of the human chromosomes is about 34 Morgans, if the S85 locus is taken to lie randomly with a uniform distribution anywhere on the human genome, the prior probability of lying in any particular stretch 0.04 Morgans long is approximately 0.04/34 = 0.0012. The prior probability of lying on another chromosome, since chromosome 16 has a length of about 1.50 Morgans, is (34.00 − 1.50)/34.00 = 0.956. The likelihood is then multiplied by the prior probability, giving the values in column five, which are relative posterior probabilities. To convert them to absolute probabilities we

Table 7.4. Calculation of probabilities assuming linkage of all families with the S85 marker on chromosome 16 (female data)

Distance from end of chromosome	Distance from S85 marker	Likelihood	Prior probability	Relative probability	Absolute probability
0.02	0.08	779	0.0012	0.935	0.012
0.06	0.04	52	0.0012	0.062	0.001
0.10	0.00	0	0.0012	0.000	0.000
0.14	0.04	52	0.0012	0.062	0.001
0.18	0.08	779	0.0012	0.935	0.012
0.22	0.12	2739	0.0012	3.287	0.041
0.26	0.26	5304	0.0012	6.364	0.079
0.30	0.20	7402	0.0012	8.883	0.111
0.34	0.24	8410	0.0012	10.092	0.126
0.38	0.28	8308	0.0012	9.970	0.124
0.42	0.32	7434	0.0012	8.291	0.111
0.46	0.36	6189	0.0012	7.427	0.093
0.50	0.40	4887	0.0012	5.865	0.073
0.54	0.44	3714	0.0012	4.456	0.056
0.58	0.48	2745	0.0012	3.294	0.041
0.62	0.52	1991	0.0012	2.839	0.030
0.66	0.56	1427	0.0012	1.712	0.021
0.70	0.60	1016	0.0012	1.219	0.015
0.74	0.64	722	0.0012	0.866	0.011
0.78	0.68	514	0.0012	0.616	0.008
0.82	0.72	367	0.0012	0.441	0.005
0.86	0.76	265	0.0012	0.371	0.004
0.90	0.80	192	0.0012	0.231	0.003
Not linked		1	0.956	0.956	0.022

divide them by their total, getting the values in column six (beyond a distance of 0.8 Morgans from the S85 locus the probabilities are small, and have been omitted). It is clear from the results that there is a good probability that the TS locus lies within 10–50 cM from the S85 locus.

However, further consideration shows that this table is unrealistic in that there seems to be a strong indication of heterogeneity. This can be investigated in a similar way, but now using two parameters, the distance from the S85 locus if linked, and the proportion α of families in which the TS locus is expected to be on chromosome 16. The calculation is done in a similar way. The distance from the end of the chromosome, and the distance from the S85 locus are calculated as in Table 7.2. The likelihood for any specific value of the distance is first calculated assuming homogeneity, with values as in Table 7.2, and then the values taking heterogeneity into account are found from Eq. 7.8. These values are shown in Table 7.2 for females and Table 7.3 for males. Unfortunately, there is a problem in assigning prior probabilities for α since we have at present little experience of heterogeneity and therefore little knowledge of what values of α occur with what frequencies. For our calculations, we have assumed as a first step that α is uniformly distributed between 0 and 1. The prior probability of finding a distance in any stretch of the chromosome 0.04 Morgans long (which we found to be 0.0012) and a value of α between 0 and 0.1, or between 0.1 and 0.2, etc., is $0.0012 \times 0.1 = 0.00012$. The likelihood is then multiplied by this, giving the relative posterior probability. Finally, these relative probabilities are converted to absolute probabilities by dividing by their total. The final results are shown in Table 7.5. The right-hand column is the total of the probabilities for a specified distance, and shows that the probability that the distance from S85 will lie in the corresponding stretch of chromosome, for example, the probability that the TS locus lies 28–32 cM from the end of the chromosome is 0.047. These probabilities show a wide spread, but for example there is a high probability that the TS locus lies in the first 40 cM from the end of the chromosome (in females). The bottom row shows the total of the corresponding column, that is, the probability distribution for α. This again has a wide spread, with the most probable values near 0.7, and a high probability of lying between 0.4 and 1.0. The probability of non-synteny is found by subtracting the total of these probabilities from 1.

It seems worth noting that these probabilities do not depend in any way on the method of sampling provided that it is effectively random. The sampling can be sequential or of fixed size, or without any predetermined plan. Unlike the significance test, the answer is the same in all cases. We do not decide whether there is or is not linkage, we give a probability that linkage is present from which appropriate conclusions can be drawn. At first sight, the method may seem to have a disadvantage that the prior probabilities are not known with any certainty.

Table 7.5. Probabilities of linkage with the S85 marker on chromosome 16 assuming heterogeneity (female data)

f	\multicolumn{10}{c}{α}	Total
	0.050	0.150	0.250	0.350	0.450	0.550	0.650	0.750	0.850	0.950	
0.02	0.000	0.000	0.002	0.005	0.012	0.020	0.026	0.025	0.015	0.004	0.109
0.06	0.000	0.000	0.002	0.008	0.018	0.030	0.036	0.030	0.014	0.002	0.141
0.10	0.000	0.001	0.004	0.012	0.027	0.043	0.048	0.033	0.011	0.000	0.179
0.14	0.000	0.000	0.002	0.008	0.018	0.030	0.036	0.030	0.014	0.002	0.141
0.18	0.000	0.000	0.002	0.005	0.012	0.020	0.026	0.025	0.015	0.004	0.109
0.22	0.000	0.000	0.001	0.003	0.008	0.014	0.019	0.020	0.014	0.005	0.083
0.26	0.000	0.000	0.001	0.002	0.005	0.009	0.013	0.015	0.012	0.006	0.063
0.30	0.000	0.000	0.000	0.001	0.003	0.006	0.009	0.011	0.010	0.006	0.047
0.34	0.000	0.000	0.000	0.001	0.002	0.004	0.006	0.008	0.008	0.006	0.035
0.38	0.000	0.000	0.000	0.001	0.001	0.002	0.004	0.005	0.006	0.005	0.025
0.42	0.000	0.000	0.000	0.000	0.001	0.002	0.003	0.004	0.004	0.004	0.018
0.46	0.000	0.000	0.000	0.000	0.001	0.001	0.002	0.002	0.003	0.003	0.013
0.50	0.000	0.000	0.000	0.000	0.000	0.001	0.001	0.002	0.002	0.003	0.009
0.54	0.000	0.000	0.000	0.000	0.000	0.000	0.001	0.001	0.002	0.002	0.006
0.58	0.000	0.000	0.000	0.000	0.000	0.000	0.001	0.001	0.001	0.001	0.004
0.62	0.000	0.000	0.000	0.000	0.000	0.000	0.000	0.001	0.001	0.001	0.003
0.66	0.000	0.000	0.000	0.000	0.000	0.000	0.000	0.000	0.001	0.001	0.002
0.70	0.000	0.000	0.000	0.000	0.000	0.000	0.000	0.000	0.000	0.000	0.001
0.74	0.000	0.000	0.000	0.000	0.000	0.000	0.000	0.000	0.000	0.000	0.001
0.78	0.000	0.000	0.000	0.000	0.000	0.000	0.000	0.000	0.000	0.000	0.001
Total	0.000	0.002	0.014	0.048	0.109	0.183	0.231	0.213	0.134	0.057	

But there is no way of circumventing this problem: methods that appear to do so will still depend on the prior distribution in some way, although it may not be at all obvious at first glance. In any case, any sensible prior distribution will have only a small quantitative effect when the amount of family or pedigree data is large.

As we have said, this analysis using only nuclear families could be considered as a first step to a full analysis using the whole data. What advantage may be gained depends on circumstances: if the original data consist almost entirely of nuclear families then not much may be expected to be gained. On the other hand, if we have a rare disease that happens to be associated with a rare allele in distant cousins this makes it probable that there has been no recombination in the chain of relatives connecting them, indicating tight linkage. But such a close association cannot be expected to occur often. Usually we can expect most of the information about linkage to come from nuclear families, so neglecting other information may not lose too much efficiency; however, only further experience can really settle this question.

Acknowledgements

We gratefully acknowledge the granting of permission to reproduce data from the Smith and Stephens (1996) paper, published in *Annals of Human Genetics*. We have also been encouraged by Professor Pawlowitzki, Professor S. Povey and others.

References

Bailey, N.T.J. *Introduction to the Mathematical Theory of Genetic Linkage.* (1961) Clarendon Press, Oxford.

Bell, J., Haldane, J.B.S. The linkage between the genes for colour-blindness and haemophilia in man. (1937) *Proc. Roy. Soc. B* **123:** 119–150.

Fisher, R.A. Inverse probability and the use of likelihood. (1934) *Proc. Camb. Phil. Soc.* **25:** 207.

Fisher, R.A. The detection of linkage with dominant abnormalities. (1935a) *Ann. Eugen.* **6:** 187–201.

Fisher, R.A. The detection of linkage with recessive abnormalities. (1935b) *Ann. Eugen.* **6:** 339–351.

Hardy, G.H. Mendelian proportions in a mixed population. (1908) *Science* **28:** 40–50.

Kosambi, D.D. The estimation of map distances from recombination fraction. (1944) *Ann. Eug.* **12:** 172–175.

Mendel, G. Versuche über Pflanzen-Hybriden. (1866) *Verhandl. d. naturforschenden vereines in Brünn* **4:** 3–47.

Mohr, J. *A Study of Linkage in Man.* (1954) Munksgaard, Copenhagen.

Morton, N.E. Sequential tests for the detection of linkage. (1955) *Am. J. Hum. Genet.* **7:** 277–318.

Morton, N.E. The detection and estimation of linkage between the genes for elliptocytosis and the Rh blood type. (1956) *Am. J. Hum. Genet.* **8:** 80–96.

Ott, J. *Analysis of Human Genetic Linkage.* 2nd edition. (1991) Johns Hopkins University Press, Baltimore.

Penrose, L.S. The detection of autosomal linkage in data which consist of pairs of brothers and sisters of unspecified parentage. (1935) *Ann. Eugen.* **6:** 133–138.

Renwick, J.H., Schulze, J. A computer program for the processing of linkage data from large pedigrees. (1961a) *Excerpta Med. Int. Congr. Ser.* **32:** E145.

Renwick, J.H., Schulze, J. Male and female recombination fractions for the nail–patella linkage in man. (1961b) *Ann. Hum. Genet.* **28:** 379–387.

Smith, C.A.B., Stephens, D.A. Estimating linkage heterogeneity. (1996) *Ann. Hum. Genet.* **60 (2):** 161–169.

Weinberg, W. Über den Nachweis der Vererbung. (1908) *Jahreshefte verein. f. vaterl. Naturk. in Würtemberg* **60:** 368–382.

Chapter 8
Tests for detecting linkage and linkage heterogeneity
Suzanne M. Leal

Introduction

There are two types of genetic heterogeneity: allelic and non-allelic heterogeneity. For allelic heterogeneity, multiple separate alleles at the same locus are
responsible for the disease phenotype. An example of a disease displaying allelic
heterogeneity is cystic fibrosis, where a large number of different mutations
within the cystic fibrosis transmembrane conductance regulator (CFTR) gene are
responsible for disease etiology (Welsch *et al.*, 1994). In the case of non-allelic
(also known as locus or linkage) heterogeneity, different individual genes are
responsible for disease etiology. Examples of disease etiology displaying locus
heterogeneity include Charcot–Marie–Tooth disease (Bird *et al.*, 1983), tuberous
sclerosis (Fryer *et al.*, 1987; Kandt *et al.*, 1992), and adult polycystic kidney
disease (APKD) (Reeders *et al.*, 1987). Locus heterogeneity implies that the
recombination fraction between the disease phenotype and genetic markers will
not be the same in all families. In the case of allelic heterogeneity the estimate of
the recombination fraction between the disease phenotype and a set of markers
will be consistent between families. Therefore, only locus heterogeneity and not
allelic heterogeneity can be detected by linkage analysis.

Parametric and nonparametric methods have been developed to test for
linkage and linkage heterogeneity. These methods, which are described and
compared, play an important role in gene mapping of both mendelian and
complex disease traits.

GENETIC MAPPING OF DISEASE GENES
ISBN 0-12-232735-7

Parametric methods

The predivided sample test

The predivided sample test (PS-test), also known as the M-test or K-test, can be used to test for heterogeneity when there are recognizable classes of families (Morton, 1956). Examples of criteria to divide a data set into classes, c, include country of origin, investigator and simplex versus multiplex families. Each family can also be considered a separate class, as in Morton's original use of this test statistic (Morton, 1956). To test whether the recombination fraction, Θ, varies between classes, the likelihood under homogeneity, H_0 is compared with the likelihood under heterogeneity, H_1. The null hypothesis, H_0 of linkage and homogeneity specifies that the recombination fraction is the same for each class:

$$\Theta_1 = \Theta_2 = ... \; \Theta_c < \tfrac{1}{2} \qquad\qquad \textbf{(Eq. 8.1)}$$

Under the alternative hypothesis, H_1, the recombination fractions for each class are not equal and can be different for every class. To test for heterogeneity, Morton (1956) introduced the likelihood ratio test:

$$\chi^2_{c-1} \;=\; 2\ln(10)[\sum_{i=1}^{c} Z_i(\hat{\Theta}_i) - Z(\hat{\Theta})] \qquad\qquad \textbf{(Eq. 8.2)}$$

where $Z_i(\hat{\Theta})$ is the maximum logarithm of odds (lod) score that occurs at a recombination fraction of Θ_i, for the ith class ($i = 1,.., c$) and $Z(\hat{\Theta})$ is the maximum LOD score that occurs at a specific value of Θ for all families combined. Asymptotically, this test statistic under the null distribution of homogeneity follows a χ^2 distribution with $c - 1$ degrees of freedom, where c is the number of classes.

In the presence of heterogeneity it is important to reliably classify a family as either of the 'linked' (the recombination fraction between the disease phenotype and marker locus is less than 1/2) or 'unlinked' ($\Theta = 1/2$) type. This is especially important in the situation when linkage results are the only source of heterogeneity classification (i.e. classification cannot be based on the phenotype of affected kindred members). In order to classify a family as either *linked* or *unlinked*, the PS-test uses information exclusively from each family to classify it. Usually, a positive value of the maximum lod score $Z_i(\hat{\Theta})$ of the ith family is used to classify a family as being of the *linked* type.

Once heterogeneity is established, it is desirable to estimate the recombination fraction in the *linked* families. In its original application, Morton (1956) applied the PS-test after he carried out analysis of genetic linkage between the loci for rhesus (Rh) and elliptocytosis (El) on 14 families. After a significant outcome of the PS-test, Morton (1956) classified the families into two groups: those that did and those that did not show linkage between Rh and El. From the four clearly *linked* families, the recombination fraction between Rh and El was estimated. This example illustrates a drawback of the PS-test: the PS-test does not use all the information in the data, because it disregards those families that appear *unlinked*. The recombination fraction for *linked* families is usually estimated from those families that are unequivocally *linked*. This will lead to the systematic under-estimation of the recombination fraction, since generally families containing a small number of recombinants, k, are classified as *linked*, whereas those families with $k \geq m/2$ (where m is the number of meioses) are classified as *unlinked*, even though they may also be of the *linked* type.

The PS-test tends to be conservative with an actual significance level smaller than or equal to the one predicted on the basis of the χ^2 distribution under most conditions (Rao *et al.*, 1978; Ott, 1983; Risch, 1988). However, it tends to be non-conservative for medium values of the true recombination fraction ($r \approx 0.2$) and large family sizes (Ott, 1983; Risch, 1988). For power, the PS-test tends to do well except in the case of admixture with loose linkage (Risch, 1988). The PS-test performs poorly when classification is by family, which may be due in part to the large number of degrees of freedom (Ott, 1983).

The B-test

The B-test (Risch, 1988) is a likelihood-ratio test based on a β distribution for the previous distribution of the recombination fraction among families (or individuals). The test statistic is constructed on the basis of the difference of the maximum log likelihoods under H_0 and H_1 and has one degree of freedom. The B-test is generally conservative with the exception of the case of a small recombination fraction and small sample size. The test is usually more powerful than the PS-test (Risch, 1988).

The admixture test

For the PS-test and B-test the alternative hypothesis is that the recombination fraction is potentially different in every family or class. In many testing situations a more credible alternative hypothesis is that there are two family types, those with linkage ($\Theta < \frac{1}{2}$) and those without linkage ($\Theta = \frac{1}{2}$) (Smith, 1961).

Smith (1963) working in a bayesian framework also considered other situations, including the circumstance where there are more than two different family types. For example, three family types: one family type with recombination fraction, $\Theta_1 < \frac{1}{2}$; the second family type with Θ_2 where $\frac{1}{2} > \Theta_2 > \Theta_1$; and a third type with $\Theta_3 = \frac{1}{2}$. Ott (1983) reformulated this approach as a likelihood ratio test. This test was coined the admixture or A-test (Hodge *et al.*, 1983; Ott, 1983). Although mentioned by Morton and Chung (1978) this test was not applied in practice until Risch and Baron (1982) investigated X-linkage and genetic heterogeneity in bipolar-related major affective disorder and Hodge *et al.* (1983) investigated heterogeneity in insulin-dependent diabetes mellitus (IDDM).

The bivariate likelihood of the ith family for the simplest situation where there are two family types, as described by Smith (1961) is given by:

$$L_i(\alpha, \Theta_1) = \alpha\, L_i(\Theta_1) + (1 - \alpha)\, Li\, (1/2) \tag{Eq. 8.3}$$

where α is the proportion of linked families at $\Theta_1 < \frac{1}{2}$ and $1 - \alpha$ is the second proportion of families, which are unlinked, $\Theta_2 = \frac{1}{2}$. To adjust the bivariate likelihood to equal 1 at $\Theta = \frac{1}{2}$ the expression is divided by $L_i(\frac{1}{2})$ which results in:

$$L_i^{*}(\alpha, \theta_1) = \alpha\, L_i^{*}(\theta_1) + (1 - \alpha) \tag{Eq. 8.4}$$

where $L_i^{*}(\Theta_1)$ is the antilog of the LOD score for the ith family. The log likelihood for all families is given by:

$$\log_e L(\alpha,\ \Theta_1) = \sum_{i=1}^{n} Log_e[L_i^{*}(\alpha,\ \Theta_1)] \tag{Eq. 8.5}$$

The test of homogeneity of linkage ($H_0 : \alpha = 1,\ \Theta_r < \frac{1}{2};\ H_1 : \alpha < 1,\ \Theta_1 < \frac{1}{2}$) is given by:

$$x^2 = 2[Log_e L(\hat{\alpha},\ \hat{\Theta}_1) - Log_e L(1, \hat{\Theta}_r)] \tag{Eq. 8.6}$$

The test asymptotically has a χ^2 distribution with one degree of freedom (one-sided test).

Construction of a support interval can help to elucidate whether a non-significant result is due to lack of power or homogeneity. A wide support interval (e.g. extending from $\alpha = 1$ to $\alpha = 0.4$) in a non-significant test result indicates insufficient power to detect heterogeneity (Mérette *et al.*, 1991).

The A-test can also be used to test for linkage and heterogeneity, where the null hypothesis is one of no linkage (H_{00}: $\Theta = \frac{1}{2}$) and the alternative hypothesis (H_1: $\alpha < 1,\ \Theta < \frac{1}{2}$) is that both linkage and heterogeneity are present. The test statistic is:

$$A = 2[Log_e L(\hat{\alpha},\ \hat{\Theta}_1)] \tag{Eq. 8.7}$$

For this situation where the null hypothesis is $\Theta = \frac{1}{2}$, Hodge *et al.* (1983) suggested that the asymptotic null distribution of the admixture test statistic was χ^2 with one degree of freedom, while Risch (1989) suggested it was χ^2 with two degrees of freedom. Faraway (1993) demonstrated that the true distribution lies between these two χ^2 distributions. The true asymptotic distribution is quite complicated, but can be approximated by the $\max(\chi_1, \chi_2)$ where χ_1 and χ_2 are independent χ_1^2 variables. Terwilliger and Ott (1994) suggest that the criterion of a likelihood ratio above 2000:1 should be used to declare that significant evidence exists for linkage in some proportion of families within the data set. The criterion of a likelihood ratio of 2000:1 is based on the likelihood ratio of 1000:1 required in a test for linkage and the allowance for a second free parameter α in the numerator of the odds ratio.

Estimates of α and Θ_1 are simply the values for α and Θ under the alternative hypothesis, H_1 where $\log L(\alpha, \Theta_1)$ is at its maximum. Unlike the PS-test, information from all families is used to estimate the probability that the ith family is of the linked type. Given good estimates of α and Θ_1, the conditional probability w that the ith family is of the 'linked' type can be calculated by:

$$w_i(\hat{\alpha}, \hat{\Theta}_1) = \frac{\hat{\alpha} L_i^*(\hat{\Theta}_1)}{\hat{\alpha} L_i^*(\hat{\Theta}_1) + 1 - \hat{\alpha}} \qquad \textbf{(Eq. 8.8)}$$

The ith family can be classified as belonging to the linked type when $w_i > \frac{1}{2}$ (Goldstein and Dillon, 1978). However, when α and Θ_1 are estimated from small samples, $w_i (\hat{\alpha}, \hat{\theta}_i) > \hat{\alpha}$ is usually a more reliable way to classify a family (Ott, 1983).

Detecting linkage and linkage heterogeneity A number of researchers have investigated the sample sizes needed to (1) detect linkage for a genetically heterogeneous trait and (2) detect heterogeneity given linkage between the disease phenotype and marker locus (Ott, 1985; Cavalli–Sforza and King, 1986; Ott, 1986; Martinez and Goldin, 1989). These investigations were carried out for mendelian disorders with a known mode of inheritance, complete penetrance, and no phenocopies.

With increasing recombination fraction and proportion of families of the 'unlinked' type, there is a corresponding increase in the number of families needed to detect linkage for a genetically heterogeneous trait. Cavalli–Sforza and King (1986) calculated analytically the number of families, n, each with s offspring needed to detect linkage using a criterion of an expected lod score of three.

For an autosomal dominant disease, when the family mating type is phase-unknown double backcross, 6.6 families ($s = 4$) are needed to detect linkage when

$\alpha = 0.90$ of the families segregate a disease locus that is linked to the marker locus at $\Theta = 0.05$. The number of families needed to detect linkage increases to $n = 86$ when $\alpha = 0.5$ and $\Theta = 0.2$. For the above example, the number of families needed in the phase-known case are reduced to $n = 4.4$ and $n = 33$, respectively.

The power to detect linkage heterogeneity given linkage between the disease phenotype and marker locus (reject the null hypothesis of linkage homogeneity) is smallest in the vicinity of the null hypothesis of $\alpha = 1$. The power to detect heterogeneity increases as α decreases until $\alpha = 0.5$, then for values of α less than 0.5 the power decreases as α decreases (Cavalli-Sforza and King, 1986; Ott, 1986; Martinez and Goldin, 1989). The number of families needed to detect heterogeneity in the presence of linkage decreases as the recombination fraction between the disease and the marker locus decreases (Cavalli-Sforza and King, 1986; Ott, 1986).

The number of families needed to detect linkage heterogeneity was calculated using a likelihood ratio criterion of 10:1. For an autosomal dominant disease when the family mating type is phase-unknown double backcross the number of families needed to detect linkage heterogeneity is $n = 263$, when $\alpha = 0.9$ and $\Theta = 0.05$. When $\Theta = 0.2$ the number of families need to detect heterogeneity increases to $n > 10\,000$. For $\Theta = 0.05$ and $\alpha = 0.5$, the number of families need to detect heterogeneity decreases to $n = 113$. However, the number of families increases to $n = 1295$ when the proportion of linked families decreases to $\alpha = 0.1$. For the phase-known situation, the number of families needed to detect heterogeneity when $\alpha = 0.9, 0.5,$ or 0.1 and $\Theta = 0.5$ are $n = 30$, $n = 14$ and $n = 110$, respectively (Cavalli-Sforza and King, 1986).

For recessive diseases, phase-unknown mating types (which are the usual cases), do not allow for separate estimation of linkage and heterogeneity unless there are at least four affected children per family or the marker is highly informative where all parental alleles are unique. For the dominant situation, heterogeneity can be detected only if the mating type is phase-known or there are at least four informative children per family (Cavalli-Sforza and King, 1986; Ott, 1986; Martinez and Goldin, 1989). It should be noted that linkage can be tested for in the presence of undetected heterogeneity, but if present the estimated recombination fraction will be inflated (Ott, 1985).

Clerget-Darpoux *et al.* (1987) examined the power and robustness of the A-test to detect linkage heterogeneity given linkage between a disease and marker locus in the genetic analysis of common disorders, where the mode of inheritance is uncertain, incomplete penetrance is present, and there is a high disease gene frequency. With decreasing penetrance the power decreases drastically and a very large sample size is required to detect linkage heterogeneity. When penetrance is low under a dominant mode of inheritance, families with unaffected parents furnish little information.

The A-test is robust to error in the specification of genetic parameters provided that the recombination fraction is not fixed a priori (i.e. when there is a candidate gene and the recombination fraction between the disease and marker locus is fixed to 0). In this situation heterogeneity can falsely be concluded. In other situations when genetic parameters are misspecified neither linkage nor linkage heterogeneity can be falsely concluded; however, the estimates of both the recombination fraction and proportion of linked families may be strongly biased. When parameter values are incorrect, there may also be a loss of power to detect linkage heterogeneity (Clerget-Darpoux *et al.*, 1987).

Martinez and Goldin (1990) investigated the power to detect linkage and heterogeneity for a disorder due to either one of two independent disease loci: locus A and B. The population disease prevalence was kept constant at 2% and different rates of heterogeneity were acquired by varying the frequency of the disease allele for the two disease loci. Monte Carlo methods were used to simulate a highly polymorphic marker locus linked ($\Theta = 0.05$) to disease locus A and unlinked to disease locus B. Martinez and Goldin (1990) examined the power of the A-test to detect linkage and heterogeneity for various modes of transmission at each disease locus (autosomal dominant or autosomal recessive), penetrance levels of disease locus (90% or 50%), ascertainment of families (e.g. at least two affected sibs in the third generation), and proportion of cases within the population due to the linked disease locus (50% or 25%). Martinez and Goldin (1990) found that the power to detect linkage is greatest when the linked disease locus has a high penetrance, the unlinked disease locus has a low penetrance, and multiplex pedigrees are sampled. When selecting multiplex pedigrees the rate of families segregating both disease loci (mixed families) increases. However, the power is more negatively influenced by a decrease in the proportion of linked families than by an increase in the number of mixed families.

Two-locus versus admixture model Goldin (1992, 1994) compared the ability of the two-locus lod score method (Lathrop and Ott, 1990) and the A-test to detect linkage for a genetically heterogeneous trait. Simulations studies were carried out using genetic models similar to those previously described by Martinez and Goldin (1990). The comparison of the two tests was carried out for a disease with a prevalence of 2% that could be caused by either one of two dominant loci, each with a penetrance of 90%. The proportion of families linked to the first locus ($\alpha = 10\%$, 25%, or 50%) and the number of affected individuals (at least two or three) required for the pedigree to be ascertained was varied. The two tests were also compared for the same conditions as above, but with a population disease prevalence of 5%, with 50% of the families linked to the first locus and at least three affected individuals required for the pedigree to be ascertained. For each of the seven models, the simulated data consisted of 50 replicates of 20 families for a total of 1000 families per model.

The power to detect linkage was approximately the same for the A-test and the two-locus lod score method. The estimates of the recombination fraction were usually slightly improved for the two-locus lod score method over the A-test; however, the estimates of the recombination fraction for the two-locus lod score method were not robust to misspecification of disease gene frequency at the two disease loci. Even in the most extreme situation tested where the disease prevalence was 5% and 25% of the families segregated both disease loci, the power and the estimation of the recombination fraction were similar for the A-test and the two-locus lod score method (Goldin, 1992; Goldin, 1994). Goldin (1992, 1994) concluded that there is no advantage in using the two-locus lod score method over the A-test to detect linkage in the presence of genetic heterogeneity.

Efficient study designs Lander and Botstein (1986) present two methods, interval mapping and simultaneous search, to capitalize on the full power of a linkage map of markers. Interval mapping and simultaneous search can easily be applied due to the dense maps of markers which are readily available (Dib *et al.*, 1996). Utilizing these methods, the number of families necessary to map a genetically heterogeneous trait can be reduced; only one-third the number of families are required. In order to detect the presence of genetic heterogeneity, only one-fifth to one-fiftieth as many families are needed (Lander and Botstein, 1986; Martinez and Goldin, 1989).

Interval mapping makes use of flanking markers of known distance to test for linkage and linkage heterogeneity. The hypothesis of linkage to two flanking markers is more demanding and therefore easier to test than linkage to a single marker. In interval mapping an entire map of linked markers can be used to test for linkage and heterogeneity by moving the disease locus through a map of linked markers of known distance. The reduction in the number of families required to detect linkage and heterogeneity is more marked for recessive than for dominant traits.

Simultaneous search involves mapping a number of disease-causing traits simultaneously. This can be accomplished by testing a 'specific ensemble' of loci (candidate genes). Intervals 1, 2, .., *k* are each tested, because they are suspected of containing a trait-causing locus (candidate gene) a priori. If all families are segregating a marker that cosegregates with a trait locus, the power to detect linkage is increased.

Heterogeneity analysis using age of onset as a covariate Mérette *et al.* (1992) extended the admixture test (Smith, 1961; Smith, 1963; Ott, 1983) to include age of onset as a covariate. The authors proposed to distinguish linked and unlinked families through the mean age of disease onset where the type 1 family members (linked families) who inherit the linked disease gene are affected at mean age μ_1, whereas in the unlinked families (genetic and non-genetic cases), the mean age of onset of is μ_2.

Including age of onset as a covariate may be useful in detecting linkage heterogeneity for diseases where there is heterogeneity of the mean age of onset. In such situations, this method may allow for the detection of linkage heterogeneity when the 'traditional' admixture test is unable to detect hetero-geneity.

The C-test

To improve the chance of detecting linkage for a genetically heterogeneous trait in data sets that consist of small pedigrees, the C-test was proposed (MacLean *et al.*, 1992). The C-test uses the sum of individual pedigrees maximum Z values (Morton, 1956). The C-test statistic is:

$$C = \sum_{i=1}^{n} \max_{\Theta} Z_i(\Theta), \qquad \textbf{(Eq. 8.9)}$$

where $Z_i(\Theta)$ is the maximum lod score for the ith family. Since the exact probabil-ity distribution under the null hypothesis is not known, simulation studies are used to determine the critical value for the C-test.

MacLean *et al.* (1992) compared the relative efficiency of the A-test and C-test ($H_{00} : \Theta = \frac{1}{2}$; $H_1 : \alpha < 1$, $\Theta < \frac{1}{2}$) for various family sizes (2–14 offspring). The authors concluded that the C-test was more efficient than the A-test for samples consisting of smaller sibships (< 12 sibs).

The two tests were compared using the asymptotic χ^2 distribution with two degrees of freedom for the A-test. This criterion was an overly stringent one, since the true null distribution of the A-test is approximately the maximum of two χ_i^2 deviates (Faraway, 1993; Faraway, 1994).

Faraway (1994) compared the two tests again, this time using simu-lation to determine the critical value for both the A-test and the C-test. He found that the efficiency of the two tests was similar. For some situations, the A-test did better than the C-test and in other cases the reverse was true. One clear disadvantage of the C-test is its inability to estimate the propor-tion of linked families and the recombination fraction between the disease and marker locus.

Nonparametric methods

Affected sib pair (ASP) method

Morton (1983) provided a test of heterogeneity based on identical by descent (IBD) scores for multiple case families and affected sib pairs. The assumption is made that in the families that are linked, linkage is complete ($\Theta = 0.0$). Morton's test for heterogeneity relates to both inter- and intra-family heterogeneity (heterogeneity due to a proportion of individuals or families who do not segregate a particular disease locus).

Chakravarti *et al.* (1987) investigated the use of IBD data for detecting linkage heterogeneity for recessive traits. The authors present a two-stage test: the data are tested for evidence of linkage and once a positive result is obtained, the data is tested for homogeneity of linkage.

The null hypothesis of no linkage ($H_0 : \Theta = 1/2$) is tested against the alternative hypothesis ($H_1 : \Theta < 1/2$) using a one-sided test. The test statistic is:

$$T = 2\sqrt{2n}(\hat{x} - 1/2) \qquad \text{(Eq. 8.10)}$$

where x is estimated by:

$$\hat{x} = (n_1 + 2n_0)/2n \qquad \text{(Eq. 8.11)}$$

where n_2, n_1, and n_0 are the number of observations where both alleles, one allele, and neither allele are IBD, respectively. T is asymptotically distributed as a standard normal distribution. Once the null hypothesis of no linkage has been rejected, the data can then be tested for homogeneity.

Under linkage, the observed number of shared alleles IBD 2, 1, 0 have the expectation $n(1-x)^2$, $2nx(1-x)$, and nx^2, respectively. The G statistic can be used to evaluate the 'goodness-of-fit' of the observed values to those expected under linkage. The test statistic is:

$$G = \frac{n_2^2}{n(1-x)^2} + \frac{n_1^2}{2nx(1-x)} + \frac{n_0^2}{nx^2} - n \qquad \text{(Eq. 8.12)}$$

which asymptotically has a χ^2 distribution with one degree of freedom. If the test is significant suggesting linkage heterogeneity the maximum likelihood estimates of the recombination fraction and the proportion of linked families are:

$$\hat{\Theta} = (1 - \sqrt{1 - 2\hat{x}_h})/2 \qquad \textbf{(Eq. 8.13)}$$

and:

$$\hat{\alpha} = (n_2 - n_0)^2 / n(n_2 - n_1 + n_0) \qquad \textbf{(Eq. 8.14)}$$

where $\hat{x}_h$ is estimated by

$$\hat{x}_h = (n_1 - 2n_0)/2(n_2 - n_0) \qquad \textbf{(Eq. 8.15)}$$

To evaluate the power and type I error for the ASP-method the authors used Monte Carlo methods. The simulation studies demonstrated that when $\Theta = 0.5$ (no linkage) the empirical type I error (probability of falsely concluding linkage) is between 5% and 8% as the number of sib pairs, n, increases from 30 to 100. When homogeneous linkage was assumed the probability of detecting linkage (power) was small when Θ was large and n was small. For example, if $\Theta = 0.3$ the empirical power of detecting linkage was 35%, 49%, and 74% when $n = 30$, 50, and 100, respectively. The power to detect linkage was over 90% when $\Theta < 0.2$ and $n > 30$. Among those simulations demonstrating linkage, the probability of falsely specifying heterogeneity was 3–7%.

The power to detect linkage heterogeneity from sib pair data was poor. The power of detecting linkage is high (> 0.97) only when $\Theta < 0.1$, the proportion of linked cases $\alpha > 0.7$ and at least $n = 30$ sib pairs are analyzed. When linkage is detected the large majority of cases are compatible with homogeneity of linkage except when $\Theta = 0.0$. The power with which heterogeneity can be detected if $\Theta > 0.1$ is $\leq 21\%$. The only situation when heterogeneity may efficiently be detected is when $\Theta = 0.0$. For example, if $n = 100$ the power of detecting heterogeneity is 86%, 93%, 90%, and 65% when $\alpha = 0.9$, 0.7, 0.5, and 0.3, respectively. However, when $n = 50$ the power is less than 80% for all values of α.

Goldin and Gershon (1988) examined the sample sizes required for the ASP-method (Blackwelder and Elston, 1985) to detect linkage for genetically heterogeneous traits. The authors were especially interested in examining the sample sizes needed to detect linkage for psychiatric disorders characterized by reduced penetrance, phenocopies, and heterogeneity. Non-parametric methods are useful in studying psychiatric disorders since modes of inheritance and correct penetrance models are usually unknown.

The authors calculated the sample sizes needed for a power of 80% to detect linkage with a type I error rate of 5% under dominant and recessive models with incomplete penetrance and allowing for a $\Theta \leq 0.1$ between the marker and disease locus.

For a disease such as schizophrenia with a relatively low population prevalence (1%), if 50% of families are linked to a marker locus, $\Theta = 0.1$, then 60 and 120 pairs are needed for recessive and dominant inheritance, respectively. For a disorder with a higher population prevalence, for example major affective disorder (prevalence 7%), the sample size increases for an autosomal dominant mode of inheritance to $n = 160$, but remains about the same for a trait with an autosomal recessive mode of inheritance. The results indicate that the ASP-method is a feasible method for detecting linkage for rare psychiatric disorders (e.g. schizophrenia or bipolar I disorder) or other rare disorders if there is one susceptibility locus that segregates within at least 50% of families in a sample. If this criterion is not met, required sample sizes may become so large that they are not feasible to obtain. It should be noted that sample size calculations were for markers that are fully informative for all ASPs. Since in most circumstances the marker loci will not be informative for every ASP, larger sample sizes will be needed to obtain the same power of detecting linkage for a heterogeneous trait.

Affected pedigree member (APM) method

Using Monte Carlo methods, Weeks and Harby (1995) examined the power of the APM-method (Weeks and Lange, 1988; Lange and Weeks, 1990; Weeks and Lange, 1992; Weeks et al., 1995) to detect linkage for a genetically heterogeneous trait. Given a known pedigree structure, the trait and marker locus (two alleles) were simulated conditional on the phenotype of the proband. The marker locus was linked at $\Theta = 0.0$ to the disease locus. The disease trait had an autosomal recessive mode of inheritance with reduced penetrance and a disease allele frequency of 0.27. The four pedigree structures consisted of a five-member nuclear family, nine- and 15-member three-generation pedigrees, and a 45-member extended pedigree. Fixed-structure sampling and sequential sampling strategies were used to ascertain a data set, which contained 300 affected individuals. To evaluate the power to detect linkage, 500 replicates were simulated for each study.

The power to detect linkage increases as the proportion of unlinked families decreases. When the percentage of unlinked families is 0%, 50%, and 80%, the power to detect linkage is $\approx$85%, 50%, and 20%, respectively. Regardless of the pedigree structure, increasing heterogeneity had approximately the same effect on the power to detect linkage (Weeks and Harby, 1995).

The APM-based test for linkage heterogeneity, NORMIX (Matise and Weeks, 1993) was applied to simulated data to evaluate the power and type I error rate of the test statistic. Data were simulated for the Duke Alzheimer families (Pericak-Vance *et al.*, 1991) under an autosomal dominant pattern of inheritance with a penetrance of 99%. The recombination fraction between the marker and the disease was varied for a total of six simulated data sets ($\Theta = 0.0$ and $\Theta = 0.05$, and $\alpha = 0.5$, 0.7, and 1.0). Linkage heterogeneity was tested for only after the detection of linkage. Evidence for linkage was accepted if $p < 0.0001$ and the hypothesis of homogeneity was rejected if $p < 0.05$.

NORMIX detected linkage heterogeneity with a power of 54–73%. The type I error rate was very high and ranged from 19–39%. The proportion of linked families was usually underestimated (Matise and Weeks, 1993).

Comparison of nonparametric methods

Davis *et al.* (1996) compared the ability of SimAPM, SimKin, SimIBD, SimISO (Davis *et al.*, 1996), APM, H–E (Haseman and Elston, 1972), ASP (Blackwelder and Elston, 1985), and extended relative pair analysis (ERPA) (Curtis and Sham, 1994) to detect linkage in the presence of linkage heterogeneity. The simulations were carried out using genetic models similar to those previously described (Martinez and Goldin, 1990; Goldin, 1992; Goldin and Weeks, 1993; Goldin, 1994). The disease trait, which has a population prevalence of 2%, could be due to either one of two dominant loci each with a penetrance of 90%. Three hetero-geneity models were examined, where the proportion of families linked to the first locus was 50% (H_{50}), 25% (H_{25}), and 10% (H_{10}). If two or more affected indi-viduals were present in the sibship (for pedigree structure see Goldin, 1992; Goldin, 1994; Davis *et al.*, 1996) the pedigree was selected to be included in the data set. For each model the simulated data consisted of 50 replicates of 20 families for a total of 1000 families per model.

All of the nonparametric statistics examined performed best for the H_{50} model (Goldin and Weeks, 1993; Davies *et al.*, 1996). For power at $p \leq 0.05$ and the disease linked to the first locus at $\Theta = 0.05$ for H_{50}, SimIBD, SimISO, H–E, and ERPA performed particularly well (power of $\approx 100\%$), while the power for ASP was slightly lower ($\approx 95\%$), the performance of SimKin was somewhat inferior (power of $\approx 85\%$), and the power for SimAPM and APM was the lowest ($\approx 55\%$). For the H_{25} model, SimKin, SimAPM, and APM performed poorly (power of $\approx 30\%$), while ASP (power of $\approx 50\%$), H–E, and ERPA (power of $\approx 65\%$) performed better, and the power for SimIBD and SimISO was the highest ($\approx 80\%$). For the H_{10} model, power was low for all tests: SimIBD, SimISO, ASP, H–E, and APM (15–20%), and SimKin, SimAPM, and ERPA (8–10%) (Davies *et al.*, 1996).

Acknowledgements

This research was funded by NIH grant HG00008.

References

Bird, T.D., Ott, J., Giblett, E.R., Chance, P.F., Sumi, S.M., Kraft, G.H. Genetic linkage evidence for heterogeneity in Charcot–Marie–Tooth neuropathy (HMSN type I). (1983) *Ann. Neurol.* **14 (6):** 679–684.

Blackwelder, W.C., Elston, R.C. A comparison of sib-pair linkage tests for disease susceptibility loci. (1985) *Genet. Epidemiol.* **2 (1):** 85–97.

Cavalli-Sforza, L.L., King, M.C. Detecting linkage for genetically heterogeneous diseases and detecting heterogeneity with linkage data. (1986) *Am. J. Hum. Genet.* **38 (5):** 599–616.

Chakravarti, A., Badner, J.A., Li, C.C. Tests of linkage and heterogeneity in Mendelian diseases using identity by descent scores. (1987) *Genet. Epidemiol.* **4 (4):** 255–266.

Clerget-Darpoux, F., Babron, M.C., Bonaiti Pellie, C. Power and robustness of the linkage homogeneity test in genetic analysis of common disorders. (1987) *J. Psychiatr. Res.* **21 (4):** 625–630.

Curtis, D., Sham, P.C. Using risk calculation to implement an extended relative pair analysis. (1994) *Ann. Hum. Genet.* **58 (2):** 151–162.

Davis, S., Schroeder, M., Goldin, L.R., Weeks, D.E. Nonparametric simulation based statistics for detecting linkage in general pedigrees. (1996) *Am. J. Hum. Genet.* **58 (4):** 867–880.

Dib, C., Faure, S., Fizames, C., Samson, D., Drouot, N., Vignal, A., Millasseau, P., Marc, S., Hazan, J., Seboun, E., Lathrop, M., Gyapay, G., Morissette, J., Weissenbach, J. A comprehensive genetic map of the human genome based on 5,264 microsatellites. (1996) *Nature* **380 (6570):** 152–154.

Faraway, J.J. Distribution of the admixture test for the detection of linkage under heterogeneity. (1993) *Genet. Epidemiol.* **10 (1):** 75–83.

Faraway, J.J. Testing for linkage under heterogeneity: A test versus C test. (1994) *Am. J. Hum. Genet.* **54 (3):** 563–564.

Fryer, A.E., Chalmers, A., Connor, J.M., Fraser, I., Povey, S., Yates, A.D., Yates, J.R., Osborne, J.P. Evidence that the gene for tuberous sclerosis is on chromosome 9. (1987) *Lancet* **1 (8534):** 659–661.

Goldin, L.R., Gershon, E.S. Power of the affected-sib-pair method for heterogeneous disorders. (1988) *Genet. Epidemiol.* **5 (1):** 35–42.

Goldin, L.R. Detection of linkage under heterogeneity: comparison of the two-locus vs. admixture models. (1992) *Genet. Epidemiol.* **9 (1):** 61–66.

Goldin, L.R., Weeks, D.E. Two-locus models of disease: comparison of likelihood and nonparametric linkage methods. (1993) *Am. J. Hum. Genet.* **53 (4):** 908–915.

Goldin, L.R. (1994) Genetic heterogeneity and other complex models: a problem for linkage detection. In Gershon, G.S., Cloninger, C.R. (eds) *Genetic Approaches to Mental Disorders.* pp. 77–87. American Psychatric Press, Washington D.C.

Goldstein, M., Dillon, W.R. *Discrete Discriminant Analysis.* (1978) John Wiley and Sons, New York.

Haseman, J.K., Elston, R.C. The investigation of linkage between a quantitative trait and a marker locus. (1972) *Behav. Genet.* **2 (1):** 3–19.

Hodge, S.E., Anderson, C.E., Neiswanger, K., Sparkes, R.S., Rimoin, D.L. The search for heterogeneity in insulin-dependent diabetes mellitus (IDDM): linkage studies, two-locus models, and genetic heterogeneity. (1983) *Am. J. Hum. Genet.* **35 (6):** 1139–1155.

Kandt, R.S., Haines, J.L., Smith, M., Northrup, H., Gardner, R.J., Short, M.P., Dumars, K., Roach, E.S., Steingold, S., Wall, S., Blanton, S.H., Flodman, P., Kwiatkowski, D.J., Jewell, A., Weber, J.L., Roses, A.D., Pericak-Vance, M.A. Linkage of an important gene locus for tuberous sclerosis to a chromosome 16 marker for polycystic kidney disease. (1992) *Nat. Genet.* **2 (1):** 37–41.

Lander, E.S., Botstein, D. Strategies for studying heterogeneous genetic traits in humans by using a linkage map of restriction fragment length polymorphisms. (1986) *Proc. Nat. Acad. Sci. USA* **83 (19):** 7353–7357.

Lange, K., Weeks, D.E. Linkage methods for identifying genetic risk factors. (1990) *World Rev. Nutr. Diet.* **63:** 236–249.

Lathrop, G.M., Ott, J. (1990) Analysis of complex disease under oligogenic models and intrafamilial heterogeneity by the LINKAGE programs. (Abstract) *Am. J. Hum. Genet.* **47 (Suppl.):** A188.

MacLean, C.J., Ploughman, L.M., Diehl, S.R., Kendler, K.S. A new test for linkage in the presence of locus heterogeneity. (1992) *Am. J. Hum. Genet.* **50 (6):** 1259–1266.

Martinez, M.M., Goldin, L.R. The detection of linkage and heterogeneity in nuclear families for complex disorders: one versus two marker loci. (1989) *Am. J. Hum. Genet.* **44 (4):** 552–559.

Martinez, M., Goldin, L.R. Power of the linkage test for a heterogeneous disorder due to two independent inherited causes: a simulation study. (1990) *Genet. Epidemiol.* **7 (3):** 219–230.

Matise, T.C., Weeks, D.E. Detecting heterogeneity with the affected-pedigree-member (APM) method. (1993) *Genet. Epidemiol.* **10 (6):** 401–406.

Mérette, C., Lehner, T., Ott, J. Interpreting nonsignificant outcomes of heterogeneity tests in gene mapping. (1991) *Am. J. Hum. Genet.* **49 (6):** 1381–1384.

Mérette, C., King, M.C., Ott, J. Heterogeneity analysis of breast cancer families by using age at onset as a covariate. (1992) *Am. J. Hum. Genet.* **50 (3):** 515–519.

Morton, N.E. The detection and estimation of linkage between the genes for elliptocytosis and the Rh blood type. (1956) *Am. J. Hum. Genet.* **8:** 80–96.

Morton, N.E., Chung, C.S. *Genetic Epidemiology.* (1978) pp. 3–11. Academic Press, New York.

Morton, N.E. An exact linkage test for multiple case families. (1983) *Hum. Hered.* **33 (4):** 244–249.

Ott, J. Linkage analysis and family classification under heterogeneity. (1983) *Ann. Hum. Genet.* **47 (4):** 311–320.

Ott, J. *Analysis of Human Genetic Linkage.* (1985) Johns Hopkins University Press, Baltimore.

Ott, J. The number of families required to detect or exclude linkage heterogeneity. (1986) *Am. J. Hum. Genet.* **39 (2):** 159–165.

Pericak-Vance, M.A., Bebout, J.L., Gaskell, P.C. Jr, Yamaoka, L.H., Hung, W.Y., Alberts, M.J., Walker, A.P., Bartlett, R.J., Haynes, C.A., Welsh, K.A., Earl, N.L.., Heyman, A., Clark, C.M., Roses, A.D. Linkage studies in familial Alzheimer disease: evidence for chromosome 19 linkage. (1991) *Am. J. Hum. Genet.* **48 (6):** 1034–1050.

Rao, D.C., Keats, B.J., Morton, N.E., Yee, S., Lew, R. Variability of human linkage data. (1978) *Am. J. Hum. Genet.* **30 (5):** 516–529.

Reeders, S.T., Breuning, M.H., Ryynanen, M.A., Wright, A.F., Davies, K.E., King, A.W., Watson, M.L., Weatherall, D.J. A study of genetic linkage heterogeneity in adult polycystic kidney disease. (1987) *Hum. Genet.* **76 (4):** 348–351.

Risch, N. A new statistical test for linkage heterogeneity. (1988) *Am. J. Hum. Genet.* **42 (2):** 353–364.

Risch, N., Baron, M. X-linkage and genetic heterogeneity in bipolar-related major affective illness: reanalysis of linkage data. (1982) *Ann. Hum. Genet.* **46 (2):** 153–166.

Risch, N. Linkage detection tests under heterogeneity. (1989) *Genet. Epidemiol.* **6 (4):** 473–480.

Smith, C.A.B. Homogeneity test for linkage data. (1961) *Proc. Sec. Int. Cong. Hum. Genet.* **1:** 212–213.

Smith, C.A.B. Testing heterogeneity of recombination fraction values in human genetics. (1963) *Ann. Hum. Genet.* **27:** 175–182.

Terwilliger, J.D., Ott, J. *Handbook of Human Genetic Linkage* (1994). Johns Hopkins University Press, Baltimore.

Weeks, D.E., Lange, K. The affected-pedigree-member method of linkage analysis. (1988) *Am. J. Hum. Genet.* **42 (2):** 315–326.

Weeks, D.E., Lange, K. A multilocus extension of the affected-pedigree-member method of linkage analysis. (1992) *Am. J. Hum. Genet.* **50 (4):** 859–868.

Weeks, D.E., Harby, L.D. The affected-pedigree-member method: power to detect linkage. (1995) *Hum. Hered.* **45 (1):** 13–24.

Weeks, D.E., Valappil, T.I., Schroeder, M., Brown, D.L. An X-linked version of the Affected-Pedigree-Member method of linkage analysis. (1995) *Hum. Hered.* **45 (1):** 25–33.

Welsch, M.J., Tsui, T.-C., Boat, T.F., Beaudet, A.L. Cystic Fibrosis (1994) In *Molecular and Metabolic Basis of Inherited Disease*, 7th edn, eds C. Scriver, A.L. Beaudet, W.E. Sly, D. Valle. McGraw-Hill, New York, pp. 3799–3866.

SECTION III

MODEL-FREE (NONPARAMETRIC) METHODS

Chapter 9
Association and haplotype sharing due to identity by descent, with an application to genetic mapping

Martin A. van der Meulen and Gerard J. te Meerman

Abstract

By tracing copies of alleles drifting through many generations and simulating recombination, the degree of genomic similarity surrounding alleles that are 'identical by descent' (IBD) can be estimated. It appears that carriers of alleles that coalesced up to 60 generations ago will share on average about 5 centiMorgans (cM) of DNA, with a standard deviation of 8 cM. This largely explains the many reported cases of large shared genomic areas surrounding a rare mutated allele.

Genomic sharing can be detected both by association between genetic markers as single alleles or haplotypes and disease. To locate genes we propose as 'haplotype sharing statistic' (HSS) the standard deviation of the length of the shared haplotype segments between all pairs. Methodological difficulties due to small founder populations are outlined. Effects of different population parameters on the power of IBD mapping are investigated by simulation.

A generally applicable design for genomic mapping is proposed in two stages: initial finding of promising genomic locations using patients from a small geographic area, with subsequent confirmation in a larger population.

Introduction

Fisher's seminal book on the genetical theory of natural selection was written in 1930. In it Fisher gives an account of the variability involved in genetic drift and deduces upper bounds for the number of copies of mutations, depending upon their age and the population growth. Two important conclusions with respect to genetic mapping through association analysis can be drawn from his work. The first is that recent mutations cannot be numerous enough in a population for associations on the scale of a larger population. For example the maximum number of copies of a single mutant allele introduced ten generations ago is with a probability 0.999 less than $40 \times$ the population growth (Fisher, 1930). If due to multifactorial inheritance, low penetrance and strong selection against mutations, only very few copies of predisposing alleles can be observed through patients, it will be very difficult to detect them by any form of association analysis. The second conclusion is that when mutations are old, they can be very numerous on the scale of a population, but they will have coalesced many generations ago and due to recombination only a small area surrounding a mutation will be conserved. This type of linkage disequilibrium has been widely observed between rare disease alleles and anonymous genetic markers, for example in cystic fibrosis (Kerem *et al.*, 1989; Maciejko *et al.*, 1989; Morral *et al.*, 1994) and Huntington's chorea (MacDonald *et al.*, 1992). This has apparently led to the expectation of many authors (e.g. Plomin *et al.*, 1994) that a very fine mesh of genetic markers at the 1 cM level or less will be required to find low-penetrance alleles by studying association between markers and disease. Recently, however, a few claims have been presented that genomic sharing surrounding a mutated allele (i.e. identical by descent (IBD)) can be detected with initial genomic screens that are much wider than 1 or 2 cMs (Houwen *et al.*, 1994). Empirical support for this idea comes from various sources. Linkage disequilibrium in the form of haplotype sharing over large genomic distances (5–15 cM) has been reported, for example melanoma (Cannon-Albright *et al.*, 1994; Gruis *et al.*, 1994), polyposis coli (Nystrom-Lahti *et al.*, 1994), benign recurrent intrahepatic cholestasis (Houwen *et al.*, 1994), diastrophic dysplasia (Hastbacka *et al.*, 1994), cartilage–hair hypoplasia (Sulisalo *et al.*, 1994), IgE (Meyers *et al.*, 1994) and the A455E cystic fibrosis mutation (De Vries *et al.*, 1996).

The purpose of this contribution is twofold:

(1) To analyse the combined effect of genetic drift and recombination with regard to genomic sharing surrounding alleles that are IBD.
(2) To present a statistical method to assist in systematic genome-wide haplotype comparison as required for IBD mapping.

116

Simulation method for genetic drift and recombination

We have shown elsewhere that the conserved area surrounding a common allele of a gene has quite a large variance, leading to a sizeable probability of extended haplotype sharing between carriers of the same disease allele (te Meerman *et al.*, 1995). Nevertheless, the empirically observed size of the shared area between apparently unrelated individuals is so large that it can only be explained as hidden consanguinity. We have investigated the process of genetic drift and recombination by following alleles descending through 60 generations (about 1500 years) as this is a realistic time frame for the introduction of alleles in many populations that have remained relatively isolated since then. Assuming a population growth of 5.5% per generation, a population increase of a factor 20 is present. Family size has been simulated by assuming a quasi-geometrical distribution according to Lotka, as cited by Feller (1957). This distribution fits the number of offspring in the American population, as cited by Feller (1957). This leads to a probability distribution with the probability to have *k* copies in the next generation $P(k)$ defined as:

$$P(k) = \frac{2 \times \alpha \times p^k}{(2 - p)^{k+1}}$$

$$P(0) = 0.56, \alpha = 0.2, p = 0.736$$

(Eq. 9.1)

This distribution has a larger variance than the Poisson distribution, which implies that genetic drift in human populations will be stronger than expected under the Poisson model used by Fisher (1930) for plants. There is an extinction probability after 60 generations of 96%, which differs little from expectation according to Fisher when the population is stationary and the number of offspring is Poisson distributed: 96.7%. Conditional on the survival of an allele, this leads to an expected number of copies of 440, due to the combined effect of drift and population growth. The distribution of the number of copies is almost exponential, as predicted by Fisher (1930).

The coalescence time is the number of generations we have to go back in order to find a common ancestor. The larger this coalescence time is, the less genomic sharing surrounding an allele IBD can be expected between carriers of an identical allele. We have statistically evaluated the coalescence time in two ways: as the time to first coalescence and as the total number of meioses connecting all observed copies (meiotic count). We have further calculated the expected size of genomic overlap between all individuals sharing the same allele IBD from the number of meioses between each pair. The results are summarized in Table 9.1.

It appears (data not shown) that the meiotic count divided by the number of

Table 9.1. Number of generations to first coalescence, meiotic count per individual and expected sharing, as function of the observed fraction.

Observed fraction (%)	Generations to first coalescence Mean (S.D)	Meiotic count per individual Mean (S.D.)	Sharing all pairs of alleles (cM) Mean (S.D.)
100	1.4 (1.2)	3.1 (0.28)	4.7 (8.4)
20	3.3 (3.8)	7.9 (1.1)	4.7 (8.4)
10	5.4 (5.7)	11.6 (1.8)	4.7 (8.3)
5	8.7 (8.0)	16.5 (3.2)	4.7 (7.9)

Simulation assumptions: 60 generations, 5.1% growth per generation, quasi-exponential distribution for the number of offspring.
10 622 simulations to find 440 replicates of surviving alleles.
Standard deviations (S.D.) computed as the mean of standard deviations within simulations.
The results are weighted for the number of copies present in the population as we expect to sample from sets of alleles proportional to their size.

copies in the observed generation is almost constant irrespective of the number of copies to which an allele actually drifted. This is also apparent from the fact that the meiotic count per individual does not change when weighted for the number of copies present and has a low standard deviation. The standard deviation for the number of generations to first coalescence is very high, within and between simulations, indicating that coalescence occurs for small subgroups of the pedigree within a few generations, and between subgroups in many generations.

The number of generations to first coalescence is directly related to the largest genomic overlap found between individuals with the same allele IBD. If two alleles coalesce in an ancestor ten generations ago, there is an expected genomic overlap of $2 \times 100/20 = 10$ cM (for a derivation see the appendix). The size of this overlap has a large standard deviation: 5.8 cM (theoretical calculation, see appendix). The expected mean sharing when weighted for the number of alleles IBD from a predecessor is almost constant (4.7 cM), but with a very high expected variance (standard deviation about 8.0 cM).

An explanation for the perhaps unexpectedly low meiotic count is the following. If 5% of all alleles is observed, the most recent common founder is present in average at generation 18.5, with a standard deviation of 13.4. The growth therefore takes place in on average 42 generations. If 5% of the alleles are observed, this amounts to a growth from 1 to $440 \times 0.05 = 22$ alleles. Assuming that the factor has been constant over all generations, the meiotic count is part of an infinite series, with multiplication factor $0.93 = (1/22)^{(1/42)}$, and as first term 22. The sum of this infinite series equals 288, which gives a meiotic count per observed individual of $288/22 = 13.1$. This is even less than what is actually observed (16.5). Assuming 60 generations of growth we expect an average meiotic count per individual of 18.9, larger than observed.

Haplotype drift–recombination equilibrium

Haplotypes inherit as rare alleles, subject to random extinction due to genetic drift and creation because of recombination. This has the consequence that although alleles generally do not disappear from the population through genetic drift, the number of observed combinations of alleles in haplotypes varies considerably compared to the number of expected combinations. This phenomenon can be described as drift–recombination equilibrium, because eventually there will be an equilibrium between haplotypes that disappear because of drift, but are also created due to recombination. We therefore agree with Kaplan *et al.* (1995) that equilibrium models are almost certainly not appropriate for rare human diseases.

The haplotype drift–recombination equilibrium can be observed by simulation of gene drop in a founder population and computing χ^2 values for the observed number of haplotypes versus the expected number on the basis of linkage equilibrium for observed allele frequencies. *Figure 9.1* shows the average value of χ^2 divided by the number of degrees of freedom as observed in a sample of 150 individuals drawn from a growing population of 1000 founders depending on the number of generations of random breeding. For a recombination frequency of 8% an equilibrium situation is reached after 16 generations, while the χ^2 value continues to increase after 50 generations for a recombination frequency of 2% and lower. These results are obtained under the same conditions as described later in detail in what is called 'the standard simulation'. In much larger populations (data not reported here), the χ^2 in a comparable sample is still higher than expected, but the effect becomes smaller. Analysis of linkage disequilibrium is therefore suitable to show the presence of founder and drift effects. Empirical support for this phenomenon comes from Peterson *et al.* (1995). When analysing haplotype sharing there is no method to discriminate between sharing due to common disease alleles and due to random IBD. The test for excess haplotype sharing will use linkage equilibrium as null hypothesis, but in human populations this assumption may not be justified. The consequence is that P values are too low.

Definition of the Haplotype Sharing Statistic

Haplotype overlap is computed starting from each marker locus between all pairs of haplotypes. As long as alleles are the same, comparison between haplotypes is continued to either side. The haplotype sharing statistic (HSS) is calculated as the standard deviation of the shared distance between haplotypes because we expect

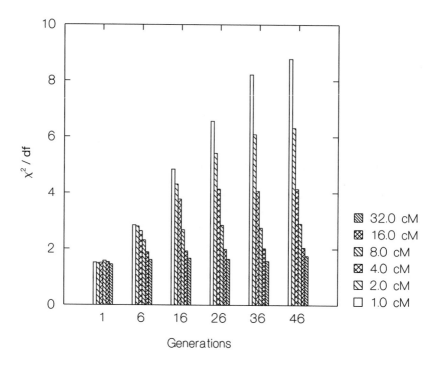

Figure 9.1 Linkage disequilibrium measured by χ^2 divided by the degrees of freedom (df) for variable number of generations of random breeding and variable recombination distances in a sample of 150 persons. The original populations had 1000 founders

that excess haplotype sharing will be observed as a few very large elements among many that are zero or almost zero. This criterion is an alternative for a previously proposed discrete criterion of sharing of three or more three-locus haplotypes (te Meerman *et al.*, 1995). The HSS includes comparison of the two independent haplotypes within a person:

$$HSS = \sqrt{\frac{\sum_{i \neq j} dist_{ij}^2 - \left[\left(\sum_{i \neq j} dist_{ij}\right)^2 / \left(N \times (N-1)\right)\right]}{\left[N \times (N-1)\right] - 1}} \quad \textbf{(Eq. 9.2)}$$

where *dist* is the calculated shared distance between two independent haplotypes, N is the number of haplotypes, and ($N * (N - 1)$) is the number of entries in the matrix of calculated shared distance between two haplotypes.

The absolute value of the HSS described above is sensitive to the degree of heterozygosity of markers and the map position of the marker on the chromosome: at the telomere we expect lower values because haplotype sharing can extend in only one direction. The HSS in the data is compared to the HSS distribution from random redistributions of the observed marker alleles over haplotypes. We have studied many distributions as they arise from multiple redistributions of observed alleles and found that a normal distribution gives a very good fit to the tail of the distribution, as can be expected according to the central limit theorem when a large set of weakly correlated stochastic variables is added. A fairly good impression of the tail of the distribution and therefore of the significance of the observed HSS in the data can already be obtained from redistributing the observed alleles a minimal number of times. Ten distributions is generally quite adequate for simulation studies where only average values are important and where a slight underestimation of statistical power is not very relevant for obtaining an impression of the effect of variations in simulation conditions. There is a correlation of 0.95 between results obtained from 10 random distributions compared to those from 100 distributions. For analysis of empirical data about 100–400 distributions are sufficient for accurate ranking of the genomic areas of interest. Because *P* values are computed using a Monte Carlo randomization test, the absolute value of the HSS is not directly relevant. The absolute amount of sharing is, however, relevant to evaluate spurious significances that may occur with minimal excess overlap. The negative common logarithm of these probabilities is useful to display the results as a kind of lod score statistic.

The computed probability of the HSS is biased due to the drift–recombination equilibrium described earlier because the random distribution of alleles causes linkage equilibrium in the Monte Carlo computation of *P* values. Therefore the probabilities cannot be interpreted absolutely. The average probability of all markers reflects the linkage disequilibrium in the population. We are, however, not interested in the absolute probability for a marker, but we are interested in the comparison between marker intervals. We achieve this by ranking of intervals between markers on the basis of the geometric mean of the two *P* values of the HSS for adjacent markers. The power of the HSS method is evaluated by ranking the probability for the area where the disease is located among other genomic areas. This is a strict criterion as it often occurs that intervals adjacent to the real position of the disease allele also have low *P* values. The rank position can be interpreted as proportional to the fraction false positive results because areas with a lower *P* value would be chosen first for confirmation. We assume that such further investigations would result in rejecting areas that do not contain the gene because what initially may be IBD appears to be identical by state (IBS). This is not necessarily sufficient, however, and investigation of another larger sample may be required to discriminate reliably between genomic regions.

Description of simulation parameters

The main properties of the standard simulation from which all variants are simulated has the following characteristics:

- 1000 non-related founders, with assigned marker haplotypes in complete linkage equilibrium, using ten alleles with frequencies proportional to ten drawings from a homogeneous distribution. The disease locus is located approximately in the middle of chromosome 1 in between two markers.
- Chromosomes are simulated with a length of 100 cM. Markers are assigned every 5, 10, 15 or 20 cM, resulting in 20, 10, 7 or 5 markers on each chromosome.
- After generating the founder population, random breeding has occurred for ten generations.
- The number of children per couple is as a truncated Poissonian distributed, with a mean of 2.25 and a maximum of ten. About 20 % of couples have no offspring.
- 20 individuals who are not related for at least the most recent four generations, are selected from the last generation. In order to be able to detect the most frequent allele at the disease locus, all alleles at the disease locus in the founder population are uniquely numbered and counted in the last generation. Ten of the 20 selected individuals carry the allele with the highest allele frequency at the disease locus, which is declared to be the disease allele. The other ten selected individuals are randomly selected without the disease allele (phenocopies). We do not evaluate the effect of using non-transmitted haplotypes as controls, which would be indicated in actual empirical investigations. For diseases that manifest themselves at advanced age, such control haplotypes would, however, need to come from spouse controls, which causes some methodological problems.

The low number of generations may seem surprising in view of the previously reported simulations with 60 generations since the introduction of the disease allele. This assumption may, however, not be too unrealistic when patients are sampled from a very small geographic area. Besides, the results can be scaled in the sense that the results can be generalized to hold for older populations provided that the map distance is reduced. The simulations reported here should primarily be seen as a methodological exercise to obtain a better understanding of the problems involved in applying haplotype sharing as a statistical method. In the analysis, only phase-known haplotypes of all markers in patients are used. To investigate the factors influencing the power of IBD mapping we have simulated variants of the standard simulation.

Results of the haplotype sharing mapping method

Data analysis

As an example of empirical data analysis, *Figure 9.2* shows results of the analysis of one simulation under standard simulation conditions with a marker spacing of 10 cM. In *Figure 9.2* the HSS in the data, the average HSS in the random distribution of observed alleles, and the final result the negative common logarithm of the P value are shown. Markers 1–10 are on chromosome 1, while the disease gene is located between markers 5 and 6 (*Figure 9.2a*). Markers 11–20 and 21–30 are on chromosomes 2 and 3, respectively (*Figures 9.2b, c*). Other chromosomes are not shown. Note the regular shape of the HSS of the random distributions over all chromosomes, which is the mean HSS of the specific marker over 100 random distributions of observed alleles. In contrast to the –log P value, the HSS in the

(a) Chromosome 1

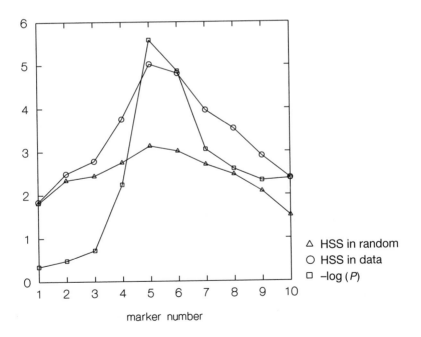

marker number

(b) Chromosome 2

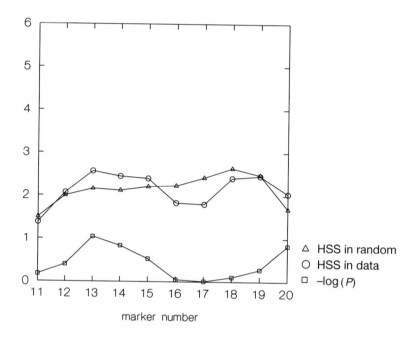

marker number

△ HSS in random
○ HSS in data
□ –log (*P*)

data and in the random distribution of alleles are not invariant for chromosomal position. *Figure 9.2* shows the results for markers, but we are particularly interested in the ranking of the interval between markers 5 and 6 on chromosome 1. This region is the most promising region.

Sensitivity of the results for changes in assumptions

Figures 9.3–9.8 show the results of the sensitivity of the HSS statistic to changes in the parameters used for simulation of the data by showing the rank position and the standard deviation of the estimate. Except for the parameter varied all parameters are as in the standard simulation. Probability values are computed from ten Monte Carlo distributions of observed alleles. Each simulation is repeated 100 times. In the figures the standard simulation is always given in open bars.

Figure 9.3 shows the results when the number of generations of random

(c)   Chromosome 3

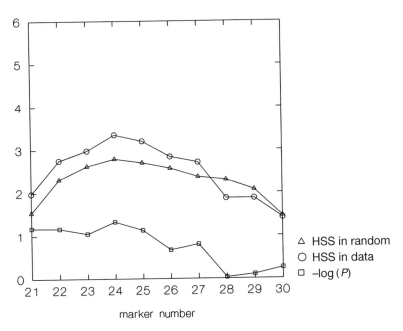

marker number

Figure 9.2 Standard deviation HSS of shared length between haplotypes for data and mean HSS for 100 random distributions in cM and the resulting –log *P* value. Results are shown for (a) markers on chromosome 1 (markers 1–10), (b) markers on chromosome 2 (markers 11–20), and (c) markers on chromosome 3 (markers 21–30). The disease gene is located between markers 5 and 6 on chromosome 1. Data from a simulation under standard conditions (see text) with marker spacing 10 cM

breeding is varied from 8–14 generations. As expected the size of the shared haplotypes decreases when the number of generations increases (te Meerman *et al.*, 1995) and therefore the rank position increases.

In *Figure 9.4* the size of the founder population is varied from 100–2000 individuals. In case of small founder populations, random drift results in alleles identical by descent with appreciable genomic sharing. Such sharing is indistinguishable from sharing due to common disease alleles except that sharing of a disease allele implies systematic overlap of haplotypes.

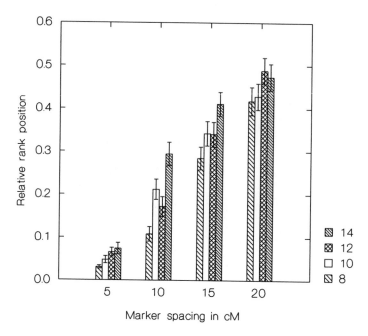

Figure 9.3 Average relative rank position and standard deviation after variable number of generations of random breeding for marker spacing of 5, 10, 15 and 20 cM

In *Figure 9.5* the homogeneity level is varied between 0% (no systematic genetic effect) to 100% (all diseased individuals have the same allele). Heterogeneity causes power to drop, for example the average ranking of the disease locus at 5 cM screen spacing deteriorates from 3% to 22% when the heterogeneity increases from 50 to 75%. When there is no gene (100% heterogeneity) the relative ranking equals 50%.

In *Figure 9.6* the number of patients used is varied from 10–40. Using more patients gives a better rank position for the disease locus.

In *Figure 9.7* the number of alleles initially present per marker is four, six and nine equiprobable alleles compared to the standard ten alleles with random frequencies. Random drift results in an exponential distribution of allele frequencies. Apparently the initial number of alleles per marker and their distribution is relatively unimportant.

In *Figure 9.8* the average number of children per couple is 2.0, 2.25 and 2.5. The effect of the rate of increase of the population on false positives is negligible.

We also investigated (not shown) the effect of the number of generations

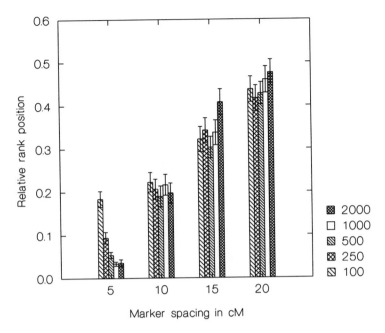

Figure 9.4 Average relative rank position and standard deviation dependent upon the number of founders of the population for marker spacing of 5, 10, 15 and 20 cM

that selected persons are not related (3–6 generations). The influence of this factor on rank position was minimal. Results from phase-unknown data are also not shown because the power drops at least an order of magnitude, even taking into account that the patient sample can then be doubled to compensate for not having to type a relative to determine phase.

Discussion

Three effects of genetic drift determine the success of haplotype mapping. Firstly is the reduction of heterogeneity by random elimination of genes predisposing to disease. Secondly genetic drift may also result in elimination of most independent descendency lines from the founder and therefore cause more recent predecessors for the disease gene(s). Thirdly haplotype drift may cause apparent excess haplotype sharing at genomic locations unrelated to the disease. The first two effects

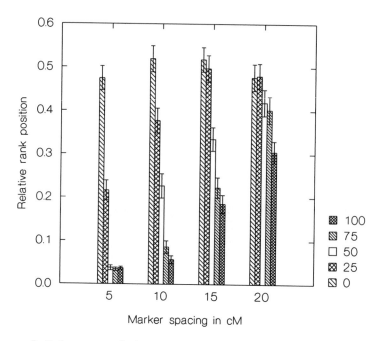

Figure 9.5 Average relative rank position and standard deviation dependent upon the homogeneity level of the disease for marker spacing of 5, 10, 15 and 20 cM

should be as strong as possible, while the third effect should be as weak as possible. The best situation can be expected for rare disease alleles in large populations, which give a high relative risk.

The expected genomic sharing and its variance between carriers of the same copy of an allele are considerable. This applies to the situation that can still be found in many places of the world where populations have not mixed much for at least 2000 years. In selecting populations to study diseases, one should concentrate on populations that have been isolated long enough and that have been growing at such a rate that genetic drift and population growth can be expected to result in high numbers of identical copies of alleles. Population growth is as important as the probability for a gene to disappear because of genetic drift because a long period of population growth may result in the presence of alleles in large numbers, even if there is a selective disadvantage.

Indications of genomic regions where a disease gene is located can be obtained by comparing all pairs of individuals for marker and haplotype sharing. Statistical analysis should be seen as complementary to genetic analysis. If

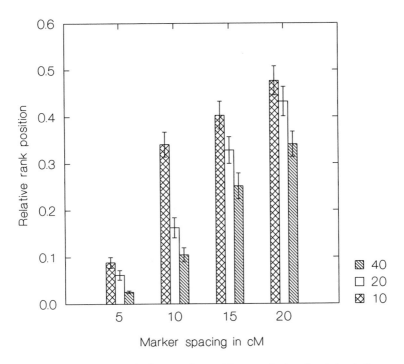

Figure 9.6 Average relative rank position and standard deviation dependent upon the number of selected patients (50% heterogeneity) for marker spacing of 5, 10, 15 and 20 cM

markers appear identical, additional investigation is sometimes required to make sure that IBD rather than by IBS is present.

The crucial aspect for statistical power is the number of alleles that is identical by descent and the number of meioses between them. This number is negatively related to the complexity of a disease (low penetrance, more genetic and environmental factors), mixture of populations, incomplete ascertainment, sampling of patients from a large geographic area and uncertain phenotypes. Increasing the number of patients from one small geographic region is very advantageous for initial mapping because geographic and genetic drift concur and lead to shorter coalescence times.

The size of the expected genomic overlap is so large that genomic screens that are presently feasible will be able to detect it, given a sufficient number of alleles that are IBD. Because the variance is so high, a strategy with an initial screen of even 10 cM could already be successful, although there will be false signals. IBD mapping is an extension of statistical mapping methodology because

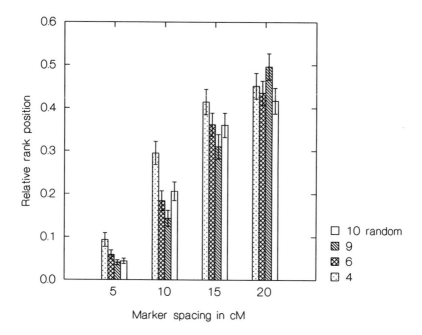

Figure 9.7 Average relative rank position and standard deviation dependent upon the number of alleles per marker: four, six and nine equiprobable alleles and the standard ten random distributed alleles for marker spacing of 5, 10, 15 and 20 cM

it offers the possibility to identify genomic regions involved in disease development using apparently unrelated individuals to whom segregation analysis cannot be applied. This method is therefore suitable for multifactorial diseases or mendelian diseases with low penetrance.

The fact that stochastic factors that cannot be controlled and can only be measured indirectly are important for the success of haplotype mapping makes the interpretation of data analysis difficult. It may be that genomic regions show up, apparently shared above chance level by affected individuals, but that subsequent verification in larger populations does not imply these regions. In founder populations genes that are marginally involved in the causation of disease may play a more important role because the population is not segregating for other risk factors.

The strategy we propose for studies using IBD mapping includes in the first step the identification of 30–100 affected individuals in a founder population from a small geographic region. Genomic screening at the 5–10 cM level will reveal a weak level of linkage disequilibrium over large genomic distances. Those genomic regions that appear to display excess sharing of haplotypes are

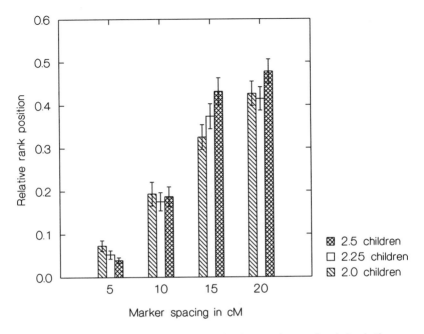

Figure 9.8 Average relative rank position and standard deviation dependent upon the average number of children per couple (2.0, 2.25 and 2.5) for marker spacing of 5, 10, 15 and 20 cM

investigated with interpolating markers. If the observed sharing appears to be due to sharing of genomic regions, a larger set of patients and controls from the surrounding population can determine if the relative risk for disease is associated with the presence of specific risk haplotypes in a genomic region. Haplotypes of patients should then show systematic overlap at only one location, pointing to a gene position. Final confirmation is obtained when gene mutations can be identified, associated with elevated risks.

Our results are obtained for single genes, but the high degree of genetic heterogeneity that is introduced makes the results for a single gene comparable to what can be expected for a gene contributing to multifactorial disease. This is particularly true when genetic drift has resulted in making multifactorial disease less multifactorial because several risk alleles of genes have drifted out the population.

Haplotype sharing analysis is to some degree comparable to association analysis with a very high level of polymorphism. The difference is, however, that haplotype sharing analysis uses all information with respect to the length of sharing, and is therefore a multilocus method.

Our results indicate that phase information contributes much to the power of haplotype sharing analysis. Consequently, the power of methods not using phase, as is the case in affected pedigree member methods should be greatly reduced compared to variants where phase is used. The reason is that sharing of large haplotypes is the rule rather than the exception, especially in large pedigrees. The most powerful method of analysis in pedigrees is of course multilocus linkage analysis using the pedigree relations between affected individuals. Such analysis is often very computing intensive. A rapid prescreening can, however, be performed with the HSS.

The actual relevance of IBD methods is not only that it offers a new perspective on the possibility of mapping disease genes using an affected-only approach, but that it also gives an additional statistical tool to analyse data obtained in other designs, most notably sib pair designs as used in asthma (Meyers *et al.*, 1994; Shirakawa *et al.*; 1994), diabetes mellitus (Davies *et al.*, 1994) and multiple sclerosis (Wood *et al.*, 1994). The emphasis is then on comparing data between affected individuals in contrast to within-sibs or small pedigrees.

With respect to the problem whether significant statistical tests give a reliable indication of the degree of involvement of a genomic region in genetic disease, we are quite pessimistic. It is clear that due to haplotype drift–recombination equilibrium computed significance levels go up considerably as the interval between markers is shortened. Statistical tests of the association type, especially those applied to small samples, will in many cases be insufficient to establish beyond doubt involvement of a genomic region in a disease. On the other hand, comparison of significance levels between genomic loci provides a rational approach for further confirmation on larger populations, although the significance levels as such are not convincing.

Acknowledgements

We are grateful to Lodewijk A. Sandkuijl for useful discussion and comments. This work is supported by the Netherlands Organisation for Scientific Research.

Appendix

The probability that no recombination occurs to the centromeric or telomeric side of an allele in N meioses at a genomic distance where the probability of recombination is t, is equal to:

$$(1 - t)^N \qquad \text{(Eq. 9.3)}$$

The corresponding probability distribution for the size of the area where no recombination occurs is:

$$N \times (1 - t)^{N-1} \qquad \text{(Eq. 9.4)}$$

The expected size of this area is, using partial integration:

$$\int_0^1 t \times N \times (1 - t)^{N-1} dt = 1/(N + 1) \qquad \text{(Eq. 9.5)}$$

The variance of the size can be computed from the integral:

$$\int_0^1 t^2 \times N \times (1 - t)^{N-1} dt - \left(\frac{1}{(N + 1)}\right)^2 \qquad \text{(Eq. 9.6)}$$

Partial integration gives as result:

$$\mathrm{var}(t) = \frac{N}{(N + 1)^2 .(N + 2)} \qquad \text{(Eq. 9.7)}$$

The expectation and variance of the addition of the telomeric and centromeric part is twice that computed above. Interference has been neglected because recombinations may take place at different meioses.

References

Cannon-Albright, L.A., Goldgar, D.E., Gruis, N.A., Neuhausen, S., Anderson, D.E., Lewis, C.M., Jost, M., Tran, T.D., Nyguen, K. Localization of the 9p melanoma susceptibility locus (MLM) to a 2-cM region between D9S736 and D9S171. (1994) *Genomics* **23 (1):** 265–268.

Davies, J.L., Kawaguchi, Y., Bennett, S.T., Copeman, J.B., Cordell, H.J., Pritchard, L.E., Reed, P.W., Gough, S.C., Jenkins, S.C., Palmer, S.M., Balfour, K.M., Rowe, B.R., Farral, M., Barnett, A.H., Bain, S.C, Todd, J.A. A genome-wide search for human type 1 diabetes susceptibility genes. (1994) *Nature* **371 (6493):** 130–136.

De Vries, H.G., van der Meulen, M.A., Rozen, R., Halley, D., Scheffer, H., ten Kate, L.P., Buys, C., te Meerman, G.J. (1996) Haplotype identity between individuals who share a CFTR mutation allele Identical By Descent: demonstration of the usefulness of the haplotype sharing concept for gene mapping in real populations. (1996) *Hum. Genet.* **98:** 304–309.

Feller, W. *An Introduction to Probability Theory and its Applications*. 2nd edition (1957) John Wiley and Sons, New York.

Fisher, R.A. *The Genetical Theory of Natural Selection*. (1930) 1st edn Clarendon Press, Dover Publications. New York. (2nd edn 1958.)

Gruis, N., Sandkuijl, L.A., Bergman, W., Frants, R.R. (1994) Genetics of the Familial Atypical Multiple mole–Melanoma Syndrome: Common 9p Haplotype in Dutch FAMMM families. PhD thesis: Leiden University.

Hastbacka, J., de la Chapelle, A., Mahtani, M.M., Clines, G., Reeve-Daly, M.P., Daly, M., Hamilton, B.A., Kusumi, K., Trivedi, B., Weaver, A., Coloma, A., Lovett, M., Buckler, A., Kartila, I., Lander, E.S. The diastrophic dysplasia gene encodes a novel sulfate transporter: positional cloning by fine-structure linkage disequilibrium mapping. (1994) *Cell* **78 (6):** 1073–1087.

Houwen, R.H., Baharloo, S., Blankenship, K., Raeymaekers, P., Juyn, J., Sandkuijl, L.A., Freimer, N.B. Genome screening by searching for shared segments: mapping a gene for benign recurrent intrahepatic cholestasis. (1994) *Nat. Genet.* **8 (4):** 380–386.

Kaplan, N.L., Hill, W.G., Weir, B.S. Likelihood methods for locating disease genes in non-equilibrium populations. (1995) *Am. J. Hum. Genet.* **56 (1):** 18–32.

Kerem, B., Rommens, J.M., Buchanan, J.A., Markiewicz, D., Cox, T.K., Chakravarti, A., Buchwald, M., Tsui, L.C. Identification of the cystic fibrosis gene: genetic analysis. (1989) *Science* **245 (4922):** 1073–1080.

MacDonald, M.E., Novelletto, A., Lin, C., Tagle, D., Barnes, G., Bates, G., Taylor, S., Allitto, B., Altherr, M., Myers, R., *et al*. The Huntington's disease candidate region exhibits many different haplotypes. (1992) *Nat. Genet.* **1 (2):** 99–103.

Maciejko, D., Bal, J., Mazurczak, T., te Meerman, G., Buys, C., Oostra, B., Halley, D. Different haplotypes for cystic fibrosis-linked DNA polymorphisms in Polish and Dutch populations. (1989) *Hum. Genet.* **83 (3):** 220–222.

Meyers, D.A., Postma, D.S., Panhuysen, C.I., Xu, J., Amelung, P.J., Levitt, R.C., Bleecker, E.R. Evidence for a locus regulating total serum IgE levels mapping to chromosome 5. (1994) *Genomics* **23 (2):** 464–470.

Morral, N., Bertranpetit, J., Estivill, X., Nunes, V., Casals, T., Gimenez, J., Reis, A., Varon Mateeva, R., Macek, M., Jr., Kalaydjieva, L., Angelicheva, D., Dancheva, K., Romeo, G., Russo, M.P., Garnerone, S., Restagno, G., Ferrari, M., Magnani, C., Claustres, M., Desgeorges, M., Schwartz, M., Novelli, G., Ferec, C., De Arce, M., Nemeti, M., Kere, J., Anvret, M., Dahl, N., Kadasi, L. The origin of the major cystic fibrosis mutation (delta F508) in European populations. (1994) *Nat. Genet.* **7 (2):** 169–175.

Nystrom–Lahti, M., Sistonen, P., Mecklin, J.P., Pylkkanen, L., Aaltonen, L.A., Jarvinen, H., Weissenbach, J., de la Chapelle, A., Peltomaki, P. Close linkage to chromosome 3p and conservation of ancestral founding haplotype in hereditary nonpolyposis colorectal cancer families. (1994) *Proc. Nat. Acad. Sci. USA* **91 (13):** 6054–6058.

Peterson, A.C., Di Rienzo, A., Lehesjoki, A.E., de la Chapelle, A., Slatkin, M., Freimer, N.B. The distribution of linkage disequilibrium over anonymous genome regions. (1995) *Hum. Mol. Genet.* **4 (5):** 887–894.

Plomin, R., Owen, M.J., McGuffin, P. The genetic basis of complex human behaviors. (1994) *Science* **264 (5166):** 1733–1739.

Puffenberger, E.G., Kauffman, E.R., Bolk, S., Matise, T.C., Washington, S.S., Angrist, M., Weissenbach, J., Garver, K.L., Mascari, M., Ladda, R., Slaugenhaupt, S.A., Chakravarti, A. Identity-by-descent and association mapping of a recessive gene for Hirschsprung disease on human chromosome 13q22. (1994) *Hum. Mol. Genet.* **3 (8):** 1217–1225.

Shirakawa, T., Li, A., Dubowitz, M., Dekker, J.W., Shaw, A.E., Faux, J.A., Ra, C., Cookson, W.O., Hopkin, J.M. Association between atopy and variants of the beta subunit of the high-affinity immunoglobulin E receptor. (1994) *Nat. Genet.* **7 (2):** 125–129.

Sulisalo, T., Francomano, C.A., Sistonen, P., Maher, J.F., McKusick, V.A., de la Chapelle, A., Kaitila, I. High-resolution genetic mapping of the cartilage–hair hypoplasia (CHH) gene in Amish and Finnish families. (1994) *Genomics* **20 (3):** 347–353.

te Meerman, G.J., van der Meulen, M.A., Sandkuijl, L.A. Perspectives of identity by descent (IBD) mapping in founder populations. (1995) *Clin. Exp. Allergy* **25 (Suppl. 2):** 97–102.

Wood, N.W., Holmans, P., Clayton, D., Robertson, N., Compston, D.A. No linkage or association between multiple sclerosis and the myelin basic protein gene in affected sibling pairs. (1994) *J. Neurol. Neurosurg. Psychiatry* **57 (10):** 1191–1194.

Chapter 10
Conditional gene identity in affected individuals

Elizabeth A. Thompson

Abstract

Gene identity by descent (IBD) underlies phenotypic similarities among relatives. It is defined relative to a given pedigree, and the prior probabilities of IBD patterns are determined by the pedigree structure. Relatives of like phenotype have, conditional on phenotype, greater than the prior probability of carrying genes IBD at loci contributing to the trait, and therefore also at linked loci. This thinking underlies all methods of linkage detection and map resolution based on affected individuals, from homozygosity mapping and sib pair methods to disequilibrium mapping.

There are two primary factors in assessing the power of a design based on sampling of affected relatives either to detect linkage or to resolve a linkage map. These are the scale of genetic distance of interest and the specificity of the disease locus. The first factor controls the probability of IBD at a linked marker at a given genetic distance, conditional on IBD at the disease locus. This is determined by the pedigree, and the greater the number of opportunities for recombination (i.e. the greater the number of ancestral segregations on paths of descent connecting the individuals), the faster the conditional probability of IBD declines with increasing genetic distance. Therefore although distant relationships provide greater power at small genetic distances and disequilibrium mapping can be used at distances smaller than can be resolved by any pedigree linkage analysis, such methods cannot detect linkage at the larger distances typical of a genome scan.

GENETIC MAPPING OF DISEASE GENES
ISBN 0-12-232735-7

The second factor controlling conditional IBD probability is the probability of IBD at the disease locus itself, conditional on the relatives being affected. For a complex disease, the specificity of any disease locus may be low among affected individuals sampled from a population. It may be increased by restricting consideration to affected members within a pedigree, but where several loci contribute jointly (e.g. multiplicatively) to disease risk, there is a bound inherent in the contribution of the locus in question to overall risk.

Specificity and scale affect all uses of affected individuals for linkage analysis, but for a given specificity and scale, information can be increased by considering several relatives jointly, or by considering several markers jointly (as in interval mapping). The conditional probabilities of multilocus multi-individual IBD patterns are not easily explicitly computed. However, the IBD pattern is a simple function of the segregation indicators in the pedigree, where an indicator for a given locus and given segregation simply indicates whether that gene is from the parent's maternal or paternal genome complement. Where explicit computation of IBD pattern probabilities is infeasible or impractical Markov chain Monte Carlo (MCMC) sampling of segregation indicators in the pedigree can provide a Monte Carlo estimate.

Introduction

Gene identity by descent (IBD) partitions the effects of genealogy from those of genetics. For a known or hypothesised model M and genealogy G, the likelihood L of M and G is the probability of observed data D which can be written as:

$$L(M,G) = P(D \mid M, G) = \sum_{B} P_M(D \mid B) \, P_G(B) \qquad \textbf{(Eq. 10.1)}$$

where the sum is over all possible states of gene IBD, B, among all the genes of all the observed individuals at all the relevant loci. In many cases, $P_M(D \mid B)$ simplifies further into the product over sets of loci of the joint probability of phenotypes of traits determined by those loci, conditional on the gene identity pattern at those loci. **Eq. 10.1** provides the basis for inference whether we are considering data on all available members of a pedigree, the affected members of a pedigree, the four haplotypes of an (affected) sib pair, the two haplotypes of an inbred affected individual, or the disease-allele-bearing chromosomes in a population. There are not different approaches to linkage analysis: there is one approach with different sampling designs appropriate to different trait models and to different ranges of values of recombination frequencies.

In genetic analysis, there is no such thing as 'model-free'. There is always the assumption that genes affect traits, and this assumption brings with it the structured model of mendelian segregation at each locus and the linear ordering of loci along a chromosome. Additionally, there are often assumptions about the population frequencies of alleles at marker loci, and implicit or explicit assumptions about the way in which trait phenotypes reflect underlying genes. There are some pedigree and sampling designs that are more robust to certain assumptions. These designs thereby lack power, when the assumptions are fulfilled, relative to less robust designs that make use of those assumptions.

Conditional gene identity by descent

Returning to gene IBD, in the context of linkage detection or of map resolution there are two papers that have been too often overlooked. The first (Edwards, 1980) addresses the sampling design appropriate for a given scale of genetic distance. We have data on disease phenotype that in some way reflect underlying patterns of gene IBD at loci contributing to the trait, and data at marker loci, which more directly reflect the descent patterns of genes at those loci. For linkage detection at recombination frequency r, it is necessary that:

$$P_r \text{ (IBD at marker locus I IBD at disease locus)} \qquad \textbf{(Pr. 10.1)}$$

should be substantially greater than the prior single-locus probability of IBD. For map resolution, the conditional probability (see **Pr. 10.1**) must be not too close to 1. Edwards (1980) defines the 'half-life' of a haplotype as the number of segregations required to bring the probability (see **Pr. 10.1**) down to 0.5. *Figure 10.1* shows the probabilities (see **Pr. 10.1**) for various different relationships or numbers of segregations. Therefore data on nuclear family relationships can detect linkage and resolve loci in the range 0.05–0.2 centiMorgans (cM). Data on more remote pedigree relationships are most useful for recombination frequencies r from 0.01–0.05. For perhaps 80 segregations relating the affected members of a large kindred or genetic isolate, the relevant scale is of the order of 10^{-3}–10^{-2} cM, while the 500 segregations in the ancestral coalescent of a sample of disease alleles of an older variant in a more extended population is useful for linkage analysis or map resolution at the scale 10^{-4}–10^{-3} cM. Note also that each class of design is appropriate over a fairly narrow range of recombination frequencies (at most one order of magnitude). There is no point doing disequilibrium mapping at the genome-search scale of $r = 0.1$ (Thompson and Neel, 1997) and no point analysing small pedigrees at the scale of $r = 10^{-3}$ (Boehnke, 1994). Further

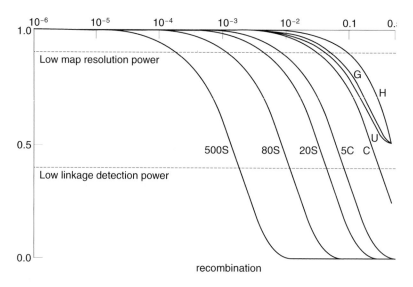

Figure 10.1 Gene identity by descent probability at a locus, conditional on identity at a locus recombination frequency r. This provides the relationship between genealogical distance, measured in numbers of intervening meioses, and genetic distance, measured in log recombination frequency. Relationships at a given genealogical distance can be used both to resolve loci and to detect linkage only within a narrow range of genetic distance. The conditional gene identity probability must be neither too close to 1, nor too small. (H, halfsib; G, grandparent; U, uncle; C, first cousin; 5C, fifth cousins; nS, n segregations)

discussion of this aspect of gene IBD in relation to disequilibrium mapping is given by Thompson and Neel (1997), and also by van der Meulen and te Meerman (see Chapter 9), and will not be pursued further here.

The second key conditional probability of gene IBD arises where we wish to map a disease locus from data on affected relatives (Bishop and Williamson, 1990). This is

P (relatives IBD at disease locus | relatives both affected),   **(Pr. 10.2)**

since the premise of all methods for linkage detection based on affected individuals is that their gene sharing at loci contributing to disease will be reflected in their genotypic similarity at linked marker loci.

For simplicity we adopt a haploid model; the results are qualitatively similar for many full diploid models for disease risk. We define risk-ratio λ_D of a susceptibility allele D with population frequency q via the penetrance probability:

$$P \text{ (affected} \mid D) = \lambda_D \, P \text{ (affected} \mid \text{not } D) = \lambda_D \, Q \text{ say, where}$$
$$P \text{ (affected)} = (q \, \lambda_D + 1 - q) \, Q \qquad \textbf{(Eq. 10.2)}$$

That is, (haploid) individuals carrying allele D have λ_D times greater probability of being affected than non-carriers.

Note that the risk-ratio for allele D as defined here is not a risk to relatives of probands as, for example, in Risch (1990); the form (see **Pr. 10.3**) will prove more useful for the current paper. Then the probability (see **Pr. 10.2**) is:

$$\frac{\pi[\lambda_D^2 q + (1-q)\,]}{\pi(\lambda_D^2 q + (1-q)) + (1-\pi)\,(\lambda_D^2 q + (1-q))^2} \qquad \textbf{(Pr. 10.3)}$$

where π is the prior probability of gene identity between the relatives in question.

Figure 10.2 shows the probabilities (see **Pr. 10.3**) for several values of π corresponding to nuclear family, near-cousin, and remote-cousin relationships. Note that if π is large, little can be gained. That is, if available, remote affected relatives provide more power to detect linkage (Bishop and Williamson, 1990); but note also the probability (see **Pr. 10.1**) then becomes relevant to the distance at which we can detect such linkage. More importantly, note the saturation with respect to λ_D, due to the non-IBD affected relatives. When λ is very large, the probability (see **Pr. 10.3**) becomes $\pi/[\pi + q\,(1-\pi)]$, which is smaller for larger q.

Analysis of complex diseases has normally been analysed in terms of penetrance of a particular allele (or, for a diploid model, genotype), the probability of being affected if carrying that allele. However, for linkage detection for such traits it is more useful to consider specificity of an allele, the probability for our simple haploid model that an affected individual carries the allele:

$$\alpha = P \text{ (}D \mid \text{affected)} = \lambda_D \, q / [\lambda_D \, q + (1 - q)] \qquad \textbf{(Eq. 10.3)}$$

Unless related affected individuals share disease alleles at a given susceptibility locus, that locus cannot be mapped by analysing marker data on affected individuals. *Figure 10.3* provides a similar result to *Figure 10.2*, but plotted in terms of specificity rather than penetrance. Plotted, for small π, is the ratio of $P(\text{IBD} \mid \text{affected})/\pi$ or:

$$\frac{\alpha^2}{q} + \frac{(1-\alpha)^2}{(1-q)} \qquad \textbf{(Eq. 10.4)}$$

141

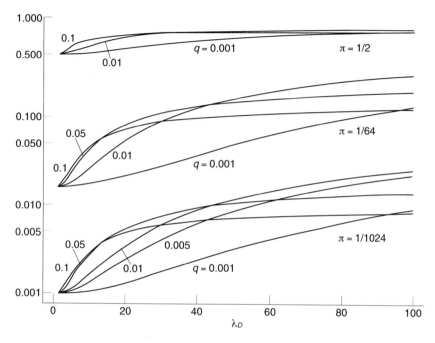

Figure 10.2 Probability of gene identity between two haplotypes at a disease locus, given two affected relatives. The probability depends on the genealogical relationship, measured by the prior gene identity probability π, the disease allele frequency q, and the risk-ratio of the disease susceptibility allele λ

We see that, for a rare allele D, the ratio of posterior to prior IBD probabilities can be greatly increased by selecting remotely related affected individuals, provided α is not too low (say $\alpha \geq 0.2$). This is as expected, however, being the case of locus heterogeneity. As has now been well established by several real-life examples, notably Alzheimer's disease (Levy-Lehad *et al.*, 1995), locus heterogeneity of a complex trait can be resolved by having several large pedigrees so that each pedigree can provide at least an indication of linkage within itself, (almost) all the affected within a pedigree being affected due to alleles at the same locus. The value of λ for such a susceptibility allele is high. A susceptibility allele, D, for a more complex trait might have a similar value of α, but a far higher population allele frequency, and correspondingly lower λ_D. For such a trait, the prospect is less favourable. Remotely related affected individuals are unlikely to share susceptibility alleles that have only moderate specificity (see *Figure 10.3*).

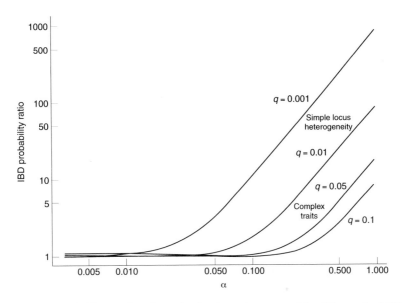

Figure 10.3 The ratio of posterior to prior gene identity probabilities in affected individuals, as a function of the specificity α of the disease susceptibility allele. The prior gene identity probability is assumed to be small

Sampling gene identity by descent

Of course, there is no clear dividing line between traits resulting from rare alleles of large effect on risk, whose genetic complexity is locus heterogeneity, and traits resulting from more frequent alleles of smaller individual effect with the same overall specificity. The ease with which linkage can be detected will depend on many factors, not least the total data set available. However, in all cases it will be desirable to have as much information from within each set of related individuals as possible, and therefore to consider jointly sets of related affected individuals.

To analyse such a pedigree design requires probabilities of gene IBD patterns, jointly over individuals and across loci. These probabilities are easily computed for any given pattern; the difficulty is the large number of alternative patterns, even at a single locus (Thompson, 1974), and therefore the huge number over linked loci. Simulation is an alternative approach. Direct simulation down the pedigree is easily and efficiently accomplished and provides realisations of multilocus multi-individual gene IBD patterns in proportion to the prior probabilities of these patterns given the pedigree structure. However, these prior probabilities bear little relation to the posterior probabilities conditional on the individuals being affected, still less conditional on data at sets of linked marker loci. To obtain good Monte Carlo estimates of these conditional probabilities, realisations from the appropriate conditional distribution must be simulated.

This conditional simulation can be accomplished by MCMC as described by Thompson (1994a). The pattern of gene IBD among a given set of individuals is a simple function of segregation indicators:

$$S_{ij} = 0 \text{ if segregation } i \text{ is grandpaternal at locus } j$$
$$= 1 \text{ if segregation } i \text{ is grandmaternal at locus } j \qquad \textbf{(Eq. 10.5)}$$

The segregation indicators, $\{S_{ij}; i=1, ..., s, j=1,.., l\}$ determine the patterns of gene identity, B, which then determine the probability of observed data (see **Eq. 10.1**). Moreover, the dependence pattern among the components of S is straightforward, with only the components S_{ij} for given j affecting the data at marker locus j, and with S_{ij} being *a priori* independent in different segregations i. Further, in the absence of interference, each S_{ij} conditionally on all the others has a simple probability distribution depending only on $S_{i,j+1}$ and $S_{i,j-1}$ the values at neighbouring loci in the same segregation (Thompson, 1994b). Hence a Metropolis algorithm can be used to sample the segregation indicators and provide accurate Monte Carlo estimates of the probabilities of gene IBD patterns conditional upon the data.

Linkage detection or linkage estimation

Realised patterns of gene IBD can be used to construct either a likelihood for linkage estimation or a test statistic for linkage detection. In the former case, realisations of B are obtained for both (hypothesised) trait and marker loci; this requires an explicit trait model. The realisations can then be used to construct a Monte Carlo estimate of the linkage likelihood or lod score curve, from which likelihood estimates of parameters, including linkage parameters, can be made (Thompson, 1994b).

An alternative is to realise patterns of gene IBD in affected individuals at marker loci conditional on marker data. This does not require any specific trait model and provides methods of linkage detection analogous to all methods based on marker similarity in affected individuals from affected pedigree member methods (Weeks and Lange, 1988), to relative pairs (Bishop and Williamson, 1990), sib pairs (Suarez *et al.*, 1978), homozygosity mapping (Lander and Botstein, 1987), and disequilibrium mapping (Hastbacka *et al.*, 1992; see also van der Meulen and te Meerman, Chapter 9). In essence, the posterior probability of shared genome at the marker loci is compared to the prior probability given only the pedigree structure providing an indication of where loci contributing to the trait may be located. To provide an effective test for linkage, the two conditional probabilities defined by **Pr. 10.1** and **Pr. 10.2** must be high. That is, the scale of recombination frequencies considered must be such that genome sharing at the markers reflects genome sharing at a trait locus (see **Pr. 10.1**), and secondly the affected individuals must, with substantial probability, carry IBD alleles at the trait locus (see **Pr. 10.2** and **Eq. 10.2**); the trait locus must have high specificity (see **Eq. 10.3**).

Conclusion

We have described how the framework of gene IBD leads to methods both for linkage estimation and linkage detection, and also facilitates assessment of the difficulties facing the linkage analysis of complex traits. The haploid model we have employed is an oversimplification of the genotypes of relatives in pedigrees, but captures many of the essential features for analyses of data on unilateral relatives in extended pedigrees or on disease-allele-bearing haplotypes. On the other hand, the model will not capture interactions between the two alleles of a genotype, such as the increased insulin-dependent diabetes mellitus (IDDM) risk of the DR3/DR4 human leucocyte antigen (HLA) genotype (Clerget-Darpoux *et al.*, 1991) or even the probability distributions exhibited by bilateral relatives, such as full sibs.

Despite these limitations, the two basic measures of conditional gene identity (see **Eq. 10.2** and **Eq. 10.3**) seem useful in assessing the power of alternative pedigree designs and sampling procedures in the detection of genes contributing to disease risk for complex traits.

References

Bishop, D.T., Williamson, J.A. The power of identity-by-state methods for linkage analysis. (1990) *Am. J. Hum. Genet.* **46 (2):** 254–265.

Boehnke, M. Limits of resolution of genetic linkage studies: implications for the positional cloning of human disease genes. (1994) *Am. J. Hum. Genet.* **55 (2):** 379–390.

Clerget-Darpoux, F., Babron, M.C., Deschamps, I., Hors, J. Complementation and maternal effect in insulin-dependent diabetes. (1991) *Am. J. Hum. Genet.* **49 (1):** 42–48.

Edwards, J.H. Allelic association in man. In Erikson, A.W. (ed.) *Population Structure and Genetic Disorders.* pp. 239–256 (1980). Acdemic Press, New York.

Hastbacka, J., de la Chapelle, A., Kaitila, I., Sistonen, P., Weaver, A., Lander, E. Linkage disequilibrium mapping in isolated founder populations: diastrophic dysplasia in Finland. (1992) *Nat. Genet.* **2 (3):** 204–211.

Lander, E.S., Botstein, D. Homozygosity mapping: a way to map human recessive traits with the DNA of inbred children. (1987) *Science* **236 (4808):** 1567–1570.

Levy-Lahad, E., Wijsman, E.M., Nemens, E., Anderson, L., Goddard, K.A., Weber, J.L., Bird, T.D., Schellenberg, G.D. A familial Alzheimer's disease locus on chromosome 1. (1995) *Science* **269 (5226):** 970–973.

Risch, N. Linkage strategies for genetically complex traits. I. Multilocus models. (1990) *Am. J. Hum. Genet.* **46 (2):** 222–228.

Suarez, B.K., Rice, J., Reich, T. The generalized sib pair IBD distribution: its use in the detection of linkage. (1978) *Ann. Hum. Genet.* **42 (1):** 87–94.

Thompson, E.A. Gene identities and multiple relationships. (1974) *Biometrics.* **30 (4):** 667–680.

Thompson, E.A. Monte Carlo estimation of multilocus autozygosity probabilities. In Sall, J., Lehman, A. (eds) *Proceedings of the 1994 Interface Conference,* pp. 498–506. (1994a) Interface Foundation of North America, Fairfax Station, VA.

Thompson, E.A. Monte Carlo likelihood in genetic mapping. (1994b) *Statistical Science* **9:** 355–366.

Thompson, E.A., Neel, J.V. (1997) Allelic association and allele frequency distribution as a function of social and demographic history. *Am. J. Hum. Genet.* **60:** 197–204.

Weeks, D.E., Lange, K. The affected-pedigree-member method of linkage analysis. (1988) *Am. J. Hum. Genet.* **42 (2):** 315–326.

Chapter 11
The affected sib pair method for linkage analysis

Michael Knapp

The aim of linkage analysis is to detect marker loci that show cosegregation with a disease of interest in pedigrees. The classical method for linkage analysis is the logarithm of odds (lod) score method (Morton, 1955; Ott, 1991). Calculation of lod scores requires specification of the mode of inheritance for the disease. Whereas this is not problematic for a disease with a simple mendelian mode of inheritance, it can become quite difficult for a genetically complex disease. By definition, a genetically complex disease is any disorder that does not follow recessive or dominant monogenic inheritance. The relation between genotype and disease phenotype can be complicated by the following phenomena.

- An individual may not become affected although he or she possesses the disease-predisposing genotype (incomplete penetrance).
- An individual without the disease-predisposing genotype may nonetheless become affected (phenocopy).
- Mutations in any one of different genes may be sufficient for the disease (locus heterogeneity).
- Mutations in each one of different genes may be necessary for the disease (oligogenic inheritance).

The presence of these complicating factors can be expected for most of the genetically influenced diseases with a high prevalence, such as asthma, diabetes mellitus or psychiatric disorders, and it is very difficult to model these factors

GENETIC MAPPING OF DISEASE GENES
ISBN 0-12-232735-7

adequately in a lod score analysis. Generally, using a wrong model will not increase the probability of obtaining a false-positive linkage result (Williamson and Amos, 1990; Amos and Williamson, 1993), but it may seriously reduce the chance of detecting a true linkage.

The advantage of the so-called non-parametric methods for linkage analysis is that they do not require specification of the mode of inheritance for the disease. These methods compare the observed agreement of alleles at a marker locus in affected individuals with the agreement expected by chance. The degree of agreement at a marker locus in two individuals can be measured either by the number of alleles identical by state (IBS) or by the number of alleles identical by descent (IBD).

Identify by state versus identity by descent

Given two individuals who have been typed at a marker locus, it is always possible to determine unambiguously if they have the same alleles at this locus. For example, both affected children in *Figure 11.1a* have allele 1, but differ in their second allele. Therefore, their IBS score (the number of alleles IBS) is 1. In order to decide if two alleles IBS are also IBD (i.e. if both alleles are copies of the same ancestral allele) it is generally required to type additional family members. For the example shown in *Figure 11.1a*, the marker genotypes of the parents reveal that the alleles 1 in the affected children are not IBD: the first child inherited its allele 1 from the mother, but the second child received an allele 1 from the father. Therefore, the IBD score is 0 for the affected sib pair in *Figure 11.1a*. Even if both parents have been typed, it is not always possible to infer the IBD score for their children (e.g. if one or both parents are homozygous). Even if both parents are heterozygous, but have the same marker genotype, one cannot decide if the IBD score for two children both having the same genotype as their parents is 0 or 2 (*Figure 11.1b*).

The affected pedigree member (APM)-method (Weeks and Lange, 1988, 1992) is a test of linkage that relies on the IBS relation. The APM-method can be used with arbitrary pedigrees. It only requires the marker genotypes of the affected individuals within the family. Other pedigree members need not to be typed. Software for calculating the test statistic for the APM-method is available from the authors of the method. In a number of recent studies, the APM-method has been applied to linkage analysis of complex diseases, for example bipolar affective disorder (Straub *et al.*, 1994), Alzheimer's disease (Schellenberg *et al.*, 1993), and schizophrenia (Wang *et al.*, 1995). A major problem with the APM-method is the fact that the population frequencies of the marker alleles must be

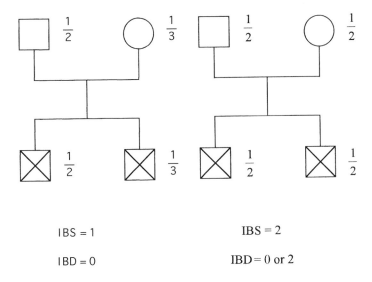

IBS = 1   IBS = 2

IBD = 0   IBD = 0 or 2

Figure 11.1 Two example marker genotype constellations for affected sib pairs

specified in order to calculate the expected number of shared marker alleles in case of no linkage. Misspecification of the allele frequencies can increase the probability of a false-positive linkage result dramatically (Babron *et al.*, 1993).

Foundations for the affected sib pair method

Affected sib pair (ASP) analysis is the most popular form of linkage analysis based on the IBD relation. Depending on the mode of inheritance for the disease, other types of affected relatives (e.g. grandparent and child) can be more powerful for the detection of linkage (Risch, 1990). However, obtaining a sample of affected sib pairs is usually less difficult than for other types of relationship.

A sample for the ASP-method consists of n nuclear families, each family consisting of two affected sibs and their parents. All family members are typed at the marker locus. Ideally, in each family, both parents possess a different heterozygous marker genotype. This assumption assures that for each sib pair the number of marker alleles IBD can be determined unambiguously. Each parent transmits each of his or her allele with probability 1/2 to a child. Therefore, the probability that two children obtain the same allele from a parent is 1/2. Therefore, two sibs will have an IBD score of 2 or 0 with probability 1/4 and an IBD score of 1 with probability 1/2.

149

If disease and marker locus are unlinked (H_0), this distribution of IBD scores is to be expected even for affected sib pairs. Let n_i denote the observed number of sib pairs sharing i marker alleles IBD ($i = 0, 1, 2$). Then, (n_2, n_1, n_0) is a realization of a trinomial distributed random variable (N_2, N_1, N_0) with parameters n and (p_2, p_1, p_0). In case of no linkage, (p_2, p_1, p_0) becomes (1/4, 1/2, 1/4).

Affected sib pair tests

Different statistical tests have been proposed to decide between the hypotheses:

$$H_0 : (p_2, p_1, p_0) = (¼, ½, ¼)$$
and $$\text{(Eq. 11.1)}$$
$$H_1 : (p_2, p_1, p_0) \neq (½, ¼, ½)$$

Blackwelder and Elston (1985) compared the power of three different tests based on the following statistics.

- Proportion of sibs sharing both alleles IBD ('two-allele' or 'proportion' test),
- Mean number of IBD alleles ('mean' test).
- A χ^2 statistic ('goodness-of-fit' test).

They concluded that although the most powerful test depends on the true state of nature (i.e. the mode of inheritance of the disease, the frequency of the disease allele, and the genetic distance between marker and disease locus), the mean test is generally more powerful than the other two tests. It can be proved that the mean test is a uniformly (in θ) most powerful test for a recessive mode of inheritance. Furthermore, the mean test is locally (in θ) most powerful irrespective of the mode of inheritance (Knapp *et al.*, 1994a).

None of the tests considered by Blackwelder and Elston (1985) make use of the genetic background of the test formulation (see **Eq. 11.1**). It can be shown (Holmans, 1993) that the vector (p_2, p_1, p_0) of IBD probabilities corresponding to a genetic model always satisfies the inequalities:

$$p_1 \leq ½ \text{ and } 2 p_0 \leq p_1 \qquad \text{(Eq. 11.2)}$$

The set of parameters for the trinomial distribution that satisfy **Eq. 11.2** and the additional trivial constraint $p_0 \geq 0$ forms a triangle in the two-dimensional plane and has therefore been called the 'possible triangle'. Holmans (1993) now proposed to test the problem (see **Eq. 11.1**) by performing a likelihood ratio test, but restricted the estimation of the IBD probabilities to this possible triangle. The null hypothesis is represented by a point at the boundary of the possible triangle. Therefore, the asymptotic distribution of this restricted likelihood ratio test is not a simple χ^2 distribution, but a mixture of χ^2 distributions with zero, one, and two degrees of freedom.

Example: Suppose that in a sample of 100 affected sib pairs, 12 sib pairs share zero, 54 share one, and 34 share two marker alleles IBD. The unrestricted maximum likelihood estimates for the ibd probabilities are $\tilde{p}_2 = 34/100$, $\tilde{p}_1 = 54/100$, and $\tilde{p}_0 = 12/100$. But these estimates do not satisfy the triangle condition (since $\tilde{p}_1 > 1/2$). It can be shown that the restricted maximum likelihood estimates are $\tilde{p}_2 = 17/46$, $\tilde{p}_1 = 1/2$, and $\tilde{p}_0 = 6/46$. The corresponding likelihood is $L_{H_1} = (17/46)^{34} \cdot (1/2)^{54} \cdot (6/46)^{12}$, whereas the likelihood under H_0 is $L_{H_0} = (1/4)^{34} \cdot (1/2)^{54} \cdot (1/4)^{12}$. Thus:

$$2 \cdot \ln (L_{H_1} / L_{H_0}) = 2 \cdot \ln \left[(34/23)^{34} \cdot (12/23)^{12}\right] = 10.9648 \qquad \textbf{(Eq. 11.3)}$$

The asymptotic P value corresponding to this value is obtained by calculating:

$$\frac{1}{2} P(\chi_1^2 > 10.9648) + \frac{\arccos \sqrt{2/3}}{2\pi} P(\chi_2^2 > 10.9648) = 0.00087 \qquad \textbf{(Eq. 11.4)}$$

Alternatively, an exact P value of 0.00082 can be calculated for the data of this example.

Instead of $2 \cdot \ln(L_{H_1}/L_{H_0})$, Holmans (1993) preferred to calculate the base 10 logarithm of the likelihood ratio (i.e. $\log_{10}(L_{H_1}/L_{H_0})$, resulting in 2.38 for the data shown above) and called this the maximum lod score. Due to the different null distributions, however, the maximum lod score value obtained by Holmans's restricted likelihood ratio test cannot be interpreted in exactly the same way as the same value in the context of classical lod score analysis.

The approach suggested by Faraway (1993, 1994) is quite similiar to Holmans's approach. He improved the usual χ^2 goodness-of-fit test (Blackwelder and Elston, 1985) by obeying the genetic restrictions (see **Eq. 11.2**) in calculating the expected proportions of sib pairs (under H_1) sharing a certain number of marker alleles IBD.

Relationship between affected sib pair tests and lod score analysis

The ASP-method has the advantage that it is not necessary to specify the mode of inheritance for the disease being studied. Now suppose that a sample of nuclear families with affected sib pairs has been obtained and classical lod score analysis is applied to these data. Performing a lod score analysis is not possible without specifying the mode of inheritance. The mode being used to perform the lod score analysis will be called the 'assumed mode' and is not necessarily identical to the mode that really governs the inheritance of the disease (the 'true mode'). Generally, ASP tests ignore the affection status of the sibs' parents. Therefore, assume that the affection status of all parents in all families is set to 'unknown' for the purpose of lod score analysis. Finally, let p' denote the frequency of the disease allele assumed for the lod score analysis. Then, Knapp *et al.* (1994b) proved that the maximum lod score Z obtained under an assumed recessive mode of inheritance is a monotone transformation of the observed total number of marker alleles $T(n) = T(n_2, n_1, n_0)$ being IBD in the sib pairs:

$$Z(n) = \begin{cases} \left| \log_{10}\left(\dfrac{2}{1+p'}\right)^{T(n)} \left(2 - \dfrac{2}{1+p'}\right)^{2n-T(n)} & \text{for} & \dfrac{2n}{1+p'} \le 2n_2 + n_1 \le 2n \\[3ex] \log_{10}\left(\dfrac{T(n)}{n}\right)^{T(n)} \left(2 - \dfrac{T(n)}{n}\right)^{2n-T(n)} & \text{for} & n \le 2n_2 + n_1 \le \dfrac{2n}{1+p'} \\[3ex] 0 & \text{for} & 2n_2 + n_1 < n \end{cases} \qquad \textbf{(Eq. 11.5)}$$

The total number of marker alleles IBD in the example given up is $T(n) = 122$. According to **Eq. 11.5**, the maximum lod score obtained for an assumed recessive mode of inheritance and an assumed disease allele frequency $p' = 0.1$ will be 2.1192. By using the LINKAGE package (Lathrop *et al.*, 1985), it can easily be checked that the maximum lod score for these data is this number obtained for $\theta = 0.24$.

From **Eq. 11.5** it follows that the mean test for ASPs is equivalent to parametric maximum lod score analysis with an assumed recessive mode of inheritance. Since the mean test is known to perform quite well over a spectrum of possible genetic mechanisms of disease, the loss of power due to falsely assuming a recessive mode of inheritance for lod score analysis is negligible if the sample consists of ASPs.

To circumvent the problem that there may be only uncertain information about the mode of inheritance for the disease, it has been proposed to maximize the lod score, not only over the recombination fraction θ, but also over the parameters specifying the mode of inheritance and the disease allele frequency (Clerget-Darpoux and Bonaïti-Pellié, 1992; MacLean *et al.*, 1993; Hodge and Elston, 1994). This has been named the MOD score analysis. For a general pedigree structure, it will be technically quite difficult and time-consuming to perform such an analysis. But for the special case of ASPs, Knapp *et al.* (1994b) showed that the MOD score is identical to Holmans's (1993) maximum lod score.

152

For the parametric lod score analysis of the data presented in the example given above, now assume a disease allele frequency of $p' = 3/14$, an additive mode of inheritance with complete penetrance of the homozygous affected genotype, and the absence of phenocopies. For this model, the maximum lod score is 2.38 ($\theta = 0.0$). Other genetic models exist for which the same score is obtained, but there is no model that leads to a larger maximum lod score than 2.38. Therefore, maximizing the maximum lod score over all single-locus disease models is identical to Holmans's base 10 logarithm of his restricted likelihood ratio.

Families with incomplete information

For ease of presentation, it has been assumed up to this point that the IBD score can be determined unambiguously for each sib pair of the sample. Since there exist no completely polymorphic markers, the sample will usually contain nuclear families in which one parent is homozygous. Inclusion of such families in Holmans's restricted likelihood ratio test is straightforward, but requires the availability of an adaquate computer program for calculating the restricted maximum likelihood estimates $\hat{p}_i$. An example of such a program is the Genetic Analysis System (GAS), which can be obtained via anonymous file transfer protocol (FTP) from `ftp.well.ox.ac.uk`.

Even sib pairs without typed parents can be analysed by Holmans's test and make a substantial contribution to the power of the analysis. As shown by Holmans (1993), a sample of $3n/2$ sib pairs without typed parents gives more power than a sample of n sib pairs with both parents being typed. Since the first sample requires $3n$ individuals to be typed, but the second sample involves typing $4n$ individuals, it may be argued that typing parents is inefficient. However, analysing sib pairs without typed parents relies on the correct specification of marker allele frequencies.

Extensions for the affected sib pair method

More than two affected sibs

Suppose that a family contains $s_k > 2$ affected sibs. From this sibship, $\binom{s_k}{2}$ different pairs of affected sibs can be formed. The IBD scores of these pairs are not mutually independent, but they are pairwise uncorrelated under the null hypo-thesis of no linkage. It follows that the proportion test and the mean test of Blackwelder and Elston (1985) are valid tests when calculated as though all pairs obtained from the same family are independent. Suarez and Van Eerdewegh (1984) proposed to give different weights to sib pairs obtained from sibships of different sizes, but it can be shown that the weights that maximize the power of the test depend on the specific alternative considered. If individuals without the disease-predisposing genotype never become affected (i.e. there are no pheno-copies), then the probability of at least one parent being homozygously affected at the disease locus (and, therefore, being uninformative for linkage) increases with an increasing number of affected children in the family. Therefore, the weights of the sib pairs should decrease with an increasing total number of affected children in the family. On the other hand, in the presence of phenocopies, the probability of a parent being homozygously unaffected at the disease locus decreases with an increasing number of affected children. In this situation, the optimal weights of sib pairs increase with an increasing total number of affected children in the family.

Inclusion of unaffected sibs

Green and Montasser (1988) proposed a sib pair test that uses the information of additional unaffected children in the families. An alternative test for this purpose can be obtained by generalizing the approach described in the preceding sub-section. Suppose that the kth family contains s_k affected and r_k unaffected children. Let $T_{AA}^{(k)}$ denote the sum of IBD scores at the marker locus for the $\binom{s_k}{2}$ pairs of affected sibs and $T_{UU}^{(k)}$ denote the sum of IBD scores at the marker locus for the $\binom{r_k}{2}$ pairs of unaffected sibs, whereas $T_{AU}^{(k)}$ gives the sum of IBD scores for the $s_k \cdot r_k$ discordant pairs. In case of linkage, $T_{AA}^{(k)}$ and $T_{UU}^{(k)}$ are expected to be larger than without linkage, whereas $T_{AU}^{(k)}$ should become smaller. Therefore, for any $c < 0$:

$$T^{(k)} = \left(T_{AA}^{(k)} - E_{H_0} T_{AA}^{(k)}\right) + c \cdot \left(T_{AU}^{(k)} - E_{H_0} T_{AU}^{(k)}\right) + c^2 \cdot \left(T_{UU}^{(k)} - E_{H_0} T_{UU}^{(k)}\right) \quad \textbf{(Eq. 11.6)}$$

is a reasonable measure for the evidence of linkage obtained by the kth family. It is straightforward to calculate the variance $Var_{H_0} T^{(k)}$ of $T^{(k)}$ in case of no linkage (Knapp *et al.*, 1995). Comparing

154

$$T := \frac{\sum_{k=1}^{n} T^{(k)}}{\sqrt{\sum_{k=1}^{n} Var_{H_o} T^{(k)}}} \qquad \text{(Eq. 11.7)}$$

with the $(1-\alpha)$ quantile of the standard normal distribution gives an approximate level α test for H_0.

The constant c is the relative weight for an unaffected sib. For $T_{UU}^{(k)}$, pairs of unaffected sibs are compared, hence this term is weighted by c^2. The weight c that maximizes the power of the test depends on the specific alternative. For a rare recessive disease, $c = -1/3$ results in a locally (in θ) most powerful test (Knapp, 1991). There are two problems with this approach. First, although some gain in power can be obtained by using the information provided by unaffected sibs, this increase in power is often quite small. An inappropriate selection of c can even reduce the power. Second, the diagnosis 'affected' is often more reliable than the diagnosis 'unaffected.' The inclusion of unaffected children will then increase the rate of misclassified individuals.

Simultaneous inclusion of two marker loci

Complex genetic diseases can be expected to involve multiple genetic factors. For parametric lod score analysis, Schork *et al.* (1993) showed that the simultaneous inclusion of two marker loci, each of them linked to one of two disease loci, can substantially increase the power in detecting linkage. In the context of the ASP-method, the results of Knapp *et al.* (1994c) indicate that a straightforward extension of the mean test for two marker loci can be much more powerful than single marker locus analysis. Cordell *et al.* (1995) extended Holmans's test to the case of two unlinked marker loci.

Conclusion

At present, the ASP-method is quite popular for locating susceptibility loci for genetically complex diseases. The genome scan for genes involved in type 2 diabetes mellitus (Hanis *et al.*, 1996) provides a very recent example. The reason most frequently cited for this popularity is that no genetic model has to be specified for ASP analysis. Considering the strong relationship between some ASP tests and parametric lod score analysis, however, it can be questioned if the labels 'model-free' and 'non-parametric' are really justified for the ASP-method (Blangero, 1995; Greenberg *et al.*, 1996). The optimal design of a linkage study will depend on a number of factors, especially on the availability of families suited for the intended type of analysis. With some diseases, it will be difficult to collect a sufficiently large number of ASPs. Therefore, the ASP-method is a valuable, but not the only tool available for complex linkage analysis.

References

Amos, C.I., Williamson, J.A. Robustness of the maximum-likelihood (LOD) method for detecting linkage. (1993) *Am. J. Hum. Genet.* **52 (1):** 213–214.

Babron, M.C., Martinez, M., Bonaïti-Pellié, C., Clerget-Darpoux, F. Linkage detection by the Affected-Pedigree-Member method: what is really tested? (1993) *Genet. Epidemiol.* **10 (6):** 389–394.

Blackwelder, W.C., Elston, R.C. A comparison of sib-pair linkage tests for disease susceptibility loci. (1985) *Genet. Epidemiol.* **2 (1):** 85–97.

Blangero, J. Genetic analysis of a common oligogenic trait with quantitative correlates: summary of GAW9 results. (1995) *Genet. Epidemiol.* **12:** 684–706.

Clerget-Darpoux, F., Bonaïti-Pellié, C. Strategies based on marker information for the study of human diseases. (1992) *Ann. Hum. Genet.* **56 (2):** 145–153.

Cordell, H.J., Todd, J.A., Bennett, S.T., Kawaguchi, Y., Farrall, M. Two-locus maximum lod score analysis of a multifactorial trait: joint consideration of IDDM2 and IDDM4 with IDDM1 in type 1 diabetes. (1995) *Am. J. Hum. Genet.* **57 (4):** 920–934.

Faraway, J.J. Improved sib-pair linkage test for disease susceptibility loci. (1993) *Genet. Epidemiol.* **10 (4):** 225–233.

Faraway, J.J. Improved sib-pair linkage test for disease susceptibility loci. (1994) *Genet. Epidemiol.* **11 (1):** 99

Green, J., Montasser, M. HLA haplotype discordance. (1988) *Biometrics.* **44 (4):** 941–950.

Greenberg, D.A., Hodge, S.E., Vieland, V.J., Spence, M.A. Affecteds only linkage methods are not a panacea. (1996) *Am. J. Hum. Genet.* **58 (4):** 892–895.

Hanis, C.L., Boerwinkle, E., Chakraborty, R., Ellsworth, D.L., Concannon, P., Stirling, B., Morrison, V.A., Wapelhorst, B., Spielman, R.S., Gogolinewens, K.J., Shephard, J.M., Williams, S.R., Risch, N., Hinds, D., Iwasaki, N., Ogata, M., Omori, Y., Petzold, C., Rietzsch, H., Schroder, H.E., Schulze, J., Cox, N.J., Menzel, S., Boriraj, V.V., Chen, X., Lim, L.R., Lindner, T., Mereu, L.E., Wang, Y.Q., Xiang, K., Yamagata, K., Yang, Y., Bell, G.I. A genome wide search for human non insulin dependent (type 2) diabetes genes reveals a major susceptibility locus on chromosome. (1996) *Nat. Genet.* **13 (2):** 161–166.

Hodge, S.E., Elston, R.C. Lods, wrods, and mods: the interpretation of lod scores calculated under different models. (1994) *Genet. Epidemiol.* **11 (4):** 329–342.

Holmans, P. Asymptotic properties of affected-sib-pair linkage analysis. (1993) *Am. J. Hum. Genet.* **52 (2):** 362–374.

Knapp, M. (1991) Statistische Methoden zur Assoziations- und Linkage-Analyse bei Kernfamilien. Rheinische Friedrich-Wilhelms Universität, Bonn.

Knapp, M., Seuchter, S.A., Baur, M.P. Linkage analysis in nuclear families. 1: Optimality criteria for affected sib-pair tests. (1994a) *Hum. Hered.* **44 (1):** 37–43.

Knapp, M., Seuchter, S.A., Baur, M.P. Linkage analysis in nuclear families. 2: Relationship between affected sib-pair tests and lod score analysis. (1994b) *Hum. Hered.* **44 (1):** 44–51.

Knapp, M., Seuchter, S.A., Baur, M.P. Two-locus disease models with two marker loci: the power of affected-sib-pair tests. (1994c) *Am. J. Hum. Genet.* **55 (5):** 1030–1041.

Knapp, M., Durner, M., Baur, M.P. Screening for linkage and association in nuclear families. (1995) *Genet. Epidemiol.* **12:** 619–623.

Lathrop, G.M., Lalouel, J.M., Julier, C., Ott, J. Multilocus linkage analysis in humans: detection of linkage and estimation of recombination. (1985) *Am. J. Hum. Genet.* **37 (3):** 482–498.

MacLean, C.J., Bishop, D.T., Sherman, S.L., Diehl, S.R. Distribution of lod scores under uncertain mode of inheritance. (1993) *Am. J. Hum. Genet.* **52 (2):** 354–361.

Morton, N.E. Sequential tests for the detection of linkage. (1955) *Am. J. Hum. Genet.* **7:** 277–318.

Ott, J. *Analysis of Human Genetic Linkage.* 2nd edition. (1991). Johns Hopkins University Press, Baltimore.

Risch, N. Linkage strategies for genetically complex traits. II. The power of affected relative pairs. (1990) *Am. J. Hum. Genet.* **46 (2):** 229–241.

Schellenberg, G.D., Payami, H., Wijsman, E.M., Orr, H.T., Goddard, K.A., Anderson, L., Nemens, E., White, J.A., Alonso, M.E., Ball, M.J., *et al.* Chromosome 14 and late-onset familial Alzheimer disease (FAD). (1993) *Am. J. Hum. Genet.* **53 (3):** 619–628.

Schork, N.J., Boehnke, M., Terwilliger, J.D., Ott, J. Two-trait-locus linkage analysis: a powerful strategy for mapping complex genetic traits. (1993) *Am. J. Hum. Genet.* **53 (5):** 1127–1136.

Straub, R.E., Lehner, T., Luo, Y., Loth, J.E., Shao, W., Sharpe, L., Alexander, J.R., Das, K., Simon, R., Fieve, R.R., *et al.* A possible vulnerability locus for bipolar affective disorder on chromosome 21q22.3. (1994) *Nat. Genet.* **8 (3):** 291–296.

Suarez, B.K., Van Eerdewegh, P. A comparison of three affected-sib-pair scoring methods to detect HLA-linked disease susceptibility genes. (1984) *Am. J. Med. Genet.* **18 (1):** 135–146.

Wang, S., Sun, C.E., Walczak, C.A., Ziegle, J.S., Kipps, B.R., Goldin, L.R., Diehl, S.R. Evidence for a susceptibility locus for schizophrenia on chromosome 6pter-p22. (1995) *Nat. Genet.* **10 (1):** 41–46.

Weeks, D.E., Lange, K. The affected-pedigree-member method of linkage analysis. (1988) *Am. J. Hum. Genet.* **42 (2):** 315–326.

Weeks, D.E., Lange, K. A multilocus extension of the affected-pedigree-member method of linkage analysis. (1992) *Am. J. Hum. Genet.* **50 (4):** 859–868.

Williamson, J.A., Amos, C.I. On the asymptotic behavior of the estimate of the recombination fraction under the null hypothesis of no linkage when the model is misspecified. (1990) *Genet. Epidemiol.* **7 (5):** 309–318.

Chapter 12
Association studies in genetic epidemiology

Max P. Baur and Michael Knapp

For the genetic dissection of complex diseases, association studies provide a complementary strategy to linkage analysis. Whereas linkage analysis addresses the question of cosegregation of two loci (no matter which alleles of the loci are present), the analysis of disease association is oriented towards overrepresentation (positive association) or underrepresentation (negative association) of a specific marker allele in diseased individuals in comparison to a control population.

A disease association can arise as a consequence of an allelic association (or linkage disequilibrium) in gametes. Let M and D denote specific alleles at two loci. If the probability of the haplotype MD (i.e. the joint occurrence of the alleles M and D) in a gamete equals the product of the probabilities for M and D at each of the loci, then M and D are said to be in linkage equilibrium. Deviations from this independent presence of M and D in haplotypes are referred as allelic association or linkage disequilibrium δ. A biological model for the occurrence of linkage disequilibrium is as follows. Suppose that in the evolutionary process, a mutation that results in allele D at the second locus occurs for the first time. In this gamete, there is also some allele (say M) at the first locus. Then, M and D are initially in complete linkage disequilibrium. In the following generations, recombination events between homologous chromosomes in the formation of gametes tend to reduce this linkage disequilibrium. The speed of this reduction depends on the genetic distance between the two loci, which is measured by the recombination fraction Θ (probability of a recombination between two loci). After n generations, the initial linkage disequilibrium is reduced by a factor of $(1 - \Theta)^n$ (Crow

GENETIC MAPPING OF DISEASE GENES
ISBN 0-12-232735-7

and Kimura, 1970). Therefore, for tightly linked loci, linkage disequilibrium can persist for hundreds of generations.

Up until the 1970s, blood group factors and the loci of the human leucocyte antigen (HLA) complex were the only polymorphisms available to be used as marker loci in disease association studies (Mourant *et al.*, 1978; Tiwari and Terasaki, 1985). The progress of molecular methodology then made it possible to examine restriction fragment length polymorphisms (RFLPs). The large majority of RFLPs are loci in the noncoding region of the DNA without any physiological significance. Therefore, disease association with a RFLP marker must rely on the presence of marked allelic association between a mutation at the unknown disease-causing locus and a certain allele of the RFLP locus. However, if the mutation rate is high, some mutations may occur by chance in gametes with one RFLP allele, but other mutations in gametes with another RFLP allele. Then, even if the RFLP is in the near vicinity of a disease-causing locus, the marker disease association study will not give a positive result.

Candidate gene disease association studies conceptually differ from marker disease association studies in that they do not rely on the presence of linkage disequilibrium. A candidate locus is a gene for which there is some a priori reason that it is itself directly involved in the pathogenesis of the disease. Candidate genes are variations at the DNA level that are supposed to result in an altered gene product or to effect the level of gene expression. Sobell *et al.* (1992) used the acronym VAPSE for 'variant affecting protein structure or expression' and described a method for generating such candidate genes.

The prior probability of observing a disease association is larger for a reasonable candidate gene than for a randomly selected anonymous marker allele. Formally, however, candidate gene disease association studies can be conceived as a special case of marker disease association studies with a recombination fraction of zero between marker and disease locus. Additionally, it should be noted that even if a candidate gene is found to be associated with a disease, this result may still be explained by the existence of a nearby disease locus being in linkage disequilibrium with the candidate gene.

Measures of association: relative risk and odds ratio

In epidemiology, the strength of an association between a risk factor R and a disease D is usually described by the relative risk (RR). If $P(D|R)$ is the probability of developing the disease for individuals with R and $P(D|\bar{R})$ is this probability for individuals without the risk factor, then the relative risk is simply the ratio of the two probabilities, that is:

$$RR = \frac{P(D \mid R)}{P(D \mid \overline{R})} \qquad \text{(Eq. 12.1)}$$

Simple transformations show that **Eq. 12.1** is identical to:

$$RR = \frac{P(R \mid D)/[(1 - P(R \mid D)]}{P(R)/[1 - P(R)]} \qquad \text{(Eq. 12.2)}$$

In genetic epidemiology, the risk factor of interest is the presence or absence of a certain allele at a marker locus. For example, Brewerton *et al.* (1973) observed that 72 of 75 patients with ankylosing spondylitis possessed at least one copy of allele B27 at the HLA-B locus, but this allele was only present in three of 75 control individuals. Therefore, 72/75 is an estimate of the prevalence of the B27-positive phenotype in individuals with ankylosing spondylitis, whereas 3/75 gives an estimate for the prevalence in the general population. (This last statement assumes that the control group was a random sample from the general population and was not selected for the absence of the disease. If the control group was a random sample from the nondiseased population, then the relative frequency of R in this group is an estimator of $P(R|\overline{D})$, which then results in an estimator of the odds ratio instead of the *RR*. See Fleiss (1973) for details and a general discussion of the concept of *RR* and odds ratio.) Inserting these estimates into **Eq. 12.2** results in an estimated *RR* of:

$$\widehat{RR} = \frac{(72/75)/(3/75)}{(3/75)/(72/75)} = 576 \qquad \text{(Eq. 12.3)}$$

It is relatively simple to perform a marker disease association study by this case–control design with two random samples of unrelated cases and controls. However, in interpreting its results, there is always the possibility that an association arises as an artefact due to population admixture. An illuminating example stems from Lander and Schork (1994): if the association between the 'trait' of the ability to eat with chopsticks and the HLA-A locus is examined in the San Francisco area, then allele A1 would be found to be positively associated with this trait simply because both the ability to use chopsticks and allele A1 is more frequent among Asians than Caucasians.

Affected family-based control association studies

To circumvent the difficulties in selecting an appropriate control group, Rubinstein *et al.* (1981) and Falk and Rubinstein (1987) proposed a design that uses parental data in place of nonrelated controls. By this approach, nuclear families with a single affected child are collected and typed at the marker locus. The two parental alleles that have not been transmitted to the affected child are combined to form the marker genotype of the 'control individual'. *Figure 12.1* gives an example for the observed marker genotypes in a nuclear family with a single affected child. In this example, the two parental alleles not inherited by the affected individual are m and m. Therefore, the genotype of the 'control individual' taken from this family is m/m . Compared to the case–control design for association studies, this method requires an increase in typing effort because three people have to be typed to obtain two 'individuals' for the sample. It can also become quite difficult to sample affected children with their parents, especially if the disease has a late age of onset. However, these disadvantages seem to be well justified by overcoming the stratification problem of the case–control design and its potential of false-positive results.

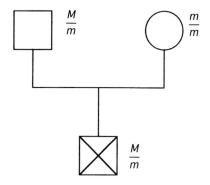

Figure 12.1 Example of marker genotypes in a nuclear family with a single affected child

Haplotype relative risk

Falk and Rubinstein (1987) claimed that the distribution of marker genotypes obtained from the nontransmitted parental alleles in families with a single affected child is identical to the distribution of marker genotypes in the population. Let W (X) denote the observed frequency of affected children who are positive (negative) for the marker allele M and let Y denote the observed frequency of families with at least one non-transmitted parental marker allele M and let Z denote the observed frequency of families with no non-transmitted parental marker allele *M*. Then, Falk and Rubinstein (1987) proposed to calculate the estimated haplotype relative risk (*HRR*):

$$HRR = \frac{W}{X} \cdot \frac{Z}{Y} \qquad \text{(Eq. 12.4)}$$

for describing the association between the marker phenotype M and the disease. Formally, there is a strong resemblance of **Eq. 12.4** to **Eq. 12.2**. It can be shown (Knapp *et al.*, 1993) that in case of $\Theta = 0$ the *HRR* obtained by using nuclear families is identical to the *RR* obtained from classical case–control studies. For $\Theta > 0$, however, the nontransmitted parental genotypes in families with a single affected child cannot be regarded as a random sample from the general distribution. For the special case of a recessive disease, this has already been noted by Ott (1989) and this result was extended to the general single locus disease model by Knapp *et al.* (1993). But even for $\Theta > 0$, the *HRR* is always at least as close to 1 as the *RR*. Therefore, even in the presence of recombinations, the *HRR* will not tend to overestimate the association between the marker and the disease.

For the construction of asymptotic confidence intervals for *HRR*, the non-independence of transmitted and nontransmitted marker phenotypes in the case of $\Theta > 0$ has to be taken into account. Let $p_{M/M}$, $p_{M/m}$, $p_{m/M}$ and $p_{m/m}$ denote the probabilities of families with a single affected child, with the first subscript denoting the marker phenotype of the affected child and with the second subscript denoting the marker phenotype of the 'control individual'. For example, $p_{M/m}$ denotes the probability of observing a nuclear family with an allele M in the affected child, but no nontransmitted parental allele M. The asymptotic variance $\sigma^2_{\widehat{HRR}}$ of $\widehat{HRR}$ is given by:

$$\left(\frac{1}{p_{./M}} + \frac{1}{p_{./m}} + \frac{1}{p_{M/.}} + \frac{1}{p_{m/.}} + 2\frac{p_{M/m} \cdot p_{m/M} - p_{M/M} \cdot p_{m/m}}{p_{./M} \cdot p_{./m} \cdot p_{M/.} \cdot p_{m/.}} \right) \cdot \frac{(HRR)^2}{n}$$

$$\text{(Eq. 12.5)}$$

with $p_{./M} = p_{M/M} + p_{m/M}$ etc. (Knapp *et al.*, 1993). By noting that:

$$\sigma_{\ln \widehat{HRR}} \approx \sigma_{\widehat{HRR}} / HRR \qquad \text{(Eq. 12.6)}$$

Eq. 12.5 can be used to construct an approximate $(1 - \alpha)$ confidence interval:

$$\left[\exp(-u_{1-\alpha/2} \cdot \sigma_{\ln \widehat{HRR}}) \cdot \widehat{HRR}, \exp(u_{1-\alpha/2} \cdot \sigma_{\ln \widehat{HRR}}) \cdot \widehat{HRR} \right] \qquad \text{(Eq. 12.7)}$$

for *HRR* ($u_{1-\alpha/2}$ denotes the $(1 - \alpha/2)$-quantile of the standard normal distribution).

Table 12.1. Observed joint distribution of transmitted and nontransmitted genotypes in a sample of nuclear families

Transmitted genotype	Nontransmitted genotype		Total
	M positive (M/M or M/m)	M negative (m/m)	
M positive (M/M or M/m)	9	14	$W = 23$
M negative (m/m)	5	1	$X = 6$
Total	$Y = 14$	$Z = 15$	29

Consider the hypothetical data shown in Table 12.1. By **Eq. 12.4**, $\widehat{HRR} =$ 4.107. Inserting $\hat{p}_{m/M} = 9/29$, $\hat{p}_{M/m} = 14/29$, $\hat{p}_{m/M} = 5/29$ and $\hat{p}_{m/m} = 1/29$ into **Eq. 12.5** gives $\sigma^2_{\widehat{HRR}} = 7.934$ and $\sigma^2_{\ln\widehat{HRR}} = 0.4703$. According to **Eq. 12.7**, an approximate 95% confidence region for *HRR* is:

$$[\exp(-1.96 \cdot 0.6858) \cdot 4.107, \exp(1.96 \cdot 0.6858) \cdot 4.107] = [1.071, 15.751]$$

(Eq. 12.8)

Due to the small sample size, the precision of the estimated *HRR* is low as indicated by this rather large confidence interval.

Testing the null hypothesis of no association

For the data of *Table 12.1*, the estimated haplotype relative risk $\widehat{HRR}$ is larger than one. Therefore, a positive association is suggested between the disease and the genotype being positive for allele M. For formally testing the null hypothesis of no association, McNemar's χ^2 test (McNemar, 1947) can be applied. Each family contributes one pair of transmitted and nontransmitted genotypes. The McNemar test uses only the information of those families in which either the transmitted or the nontransmitted genotype is positive for allele M, but the other genotype is negative for M. For the families shown in *Table 12.1*, the transmitted genotype is M positive and the nontransmitted genotype is M negative in 14 families, whereas in only five families, the transmitted genotype is negative for M and the nontransmitted genotype is M positive. In case of no association, both of these numbers are expected to be roughly the same. The statistic of McNemar's test for this example is $(14 - 5)^2/(14 + 5)$. By comparing this value with the χ^2 distribution with one degree of freedom, a P value of 0.039 is obtained. Therefore, for a prescribed α level of 0.05, the observed difference is statistically significant.

The McNemar test obeys the matching of cases and controls. Falk and Rubinstein (1987) originally proposed to compare the observed frequencies of the M positive genotype in transmitted and nontransmitted genotypes by means of the usual χ^2 test. For this approach, the matrix in *Table 12.1* is decomposed into its row and column marginals. The result is shown in *Table 12.2*.

The test statistic of the χ^2 test for *Table 12.2* is $\chi^2 = 6.046$ ($P = 0.014$). But this approach is only valid in cases where the transmitted and nontransmitted genotypes in a family are independent under H_0. A set of sufficient assumptions leading to this independence was given by Knapp *et al.* (1993). In case of population heterogeneity, however, transmitted and nontransmitted genotypes in nuclear families are not independent. Since the method that uses nuclear families for association studies was designed to overcome the problems due to population heterogeneity, it seems unreasonable to assume population homogeneity in its analysis.

Table 12.2. Observed frequencies of genotypes in transmitted and nontransmitted genotypes

	Genotype		
	M positive	M negative	Total
Cases	23	6	29
Controls	14	15	29
Total	37	21	58

Up to this point, the methods do not differentiate between genotypes being homozygous and heterozygous for allele M. Therefore, they may be viewed as methods of genotype-based analysis (Schaid and Sommer, 1994). The unit of observation is a nuclear family. Alternatively, Terwilliger and Ott (1992) proposed to classify each single parent according to his or her transmitted and nontransmitted allele. *Table 12.3* gives an example.

In case of no association, each heterozygous parent transmits its M allele to the affected child with probability 1/2. Again, McNemar's χ^2 statistic can be used to test this null hypothesis. Spielman *et al.* (1993) called this approach the transmission/disequilibrium test (TDT). For the example of *Table 12.3*, $\chi^2 = (19 - 6)^2/25 = 6.76$ ($P = 0.0093$).

Table 12.3. Observed joint distribution of transmitted and nontransmitted alleles for the parents in a sample of nuclear families

Transmitted allele	Nontransmitted allele		Total
	M	m	
M	33	19	52
m	6	0	6
Total	39	19	58

Terwilliger and Ott (1992) observed that the unmatched analysis of *Table 12.3*, which compares the observed frequencies of allele M in the group of transmitted and nontransmitted alleles (*Table 12.4*) by means of the usual χ^2 test is often more powerful than all the other tests of linkage equilibrium. They called this the haplotype-based haplotype relative risk (HHRR) test. This test, however, is also the least robust test. It requires that the contributions of the parents are independent and also that transmitted and nontransmitted genotypes are independent.

Table 12.4. Observed frequencies of alleles in transmitted and nontransmitted alleles ($\chi^2 = 8.617$, $P = 0.003$)

	Parental alleles		Total
	M	m	
Transmitted	52	6	58
Nontransmitted	39	19	58
Total	91	25	116

Sample size determination

In planning an association study, it is essential to estimate the number of nuclear families required to detect an existing association. For this, it is necessary to calculate the power of the statistical test used to decide between H_0 (no association present) and H_1 (association present). The power of a statistical test is the probability of rejecting the null hypothesis when it is false. Given the sample size and the significance level α, this power depends on the magnitude of the association between marker allele and disease. Therefore, power calculations need to specify the alternative of interest (i.e. the true state of nature that is intended to be detected by the study). Technically, a power calculation can be performed either by simulations or by asymptotic approximations.

As an example, this section presents power calculations for the transmission/disequilibrium test (TDT). For ease of presentation, the following two assumptions are made (but see the remarks at the end of this section).

- The recombination fraction Θ between marker and disease allele is zero.
- The sample is drawn from a homogenous population for which there is random mating with respect to marker genotype.

The true state of nature can then be described by specification of the relative risks Ψ_2 and Ψ_1 and the population frequency p of allele M. Ψ_2 and Ψ_1 measure the increase in disease probability for a homozygous MM and a heterozygous Mm individual, respectively, compared with the disease probability for a homozygous mm individual. Instead of the allele frequency p, it may be easier to specify the percentage of disease prevalence attributable to allele M (Schaid and Sommer, 1993), that is, the attributable risk (AR). The relation between AR and allele frequency is given by ($q = 1 - p$):

$$AR = \frac{p^2(\Psi_2 - 1) + 2pq(\Psi_1 - 1)}{p^2\Psi_2 + 2pq\Psi_1 + q^2} \qquad \textbf{(Eq. 12.9)}$$

For various combinations of Ψ_2, Ψ_1, AR and $\alpha = 0.05, 0.01, 0.001, 0.0001$, *Table 12.5* gives the sample sizes necessary to obtain a power of 90% for the TDT. Of course, the smaller the association to be detected the larger the necessary sample size. Therefore, the figures in *Table 12.5* increase with decreasing relative risks, AR, and significance level. What seems to be quite surprising are the large sample sizes required for a recessive disease model. By comparing with *Table 2* of Knapp *et al.* (1995), it can be seen that the conditional on parental genotype (CPG) test (Schaid and Sommer, 1993) is more powerful than the TDT for a recessive disease.

Table 12.5. Sample sizes to achieve 90% power

AR	α	Dominant model $\Psi_1 = \Psi_2 =$					Recessive model $\Psi_1 = 1, \Psi_2 =$					Additive model $\Psi_1 = (\Psi_2+1)/2, \Psi_2 =$				
		2	4	10	30	100	2	4	10	30	100	2	4	10	30	100
1%	.0001	7986	4221	2864	2348	2188	31344	18048	10840	6640	4246	13478	6089	3547	2579	2246
	.001	6234	3282	2238	1777	1618	24538	14069	8515	5217	3324	10528	4746	2761	1983	1683
	.01	4444	2303	1558	1238	1125	17531	10077	6077	3732	2403	7478	3356	1931	1392	1181
	.05	3105	1580	1095	827	724	12408	7173	4334	2649	1722	5296	2372	1335	963	766
5%	.0001	1705	860	575	467	438	3443	1903	1175	761	535	2865	1250	718	518	448
	.001	1333	675	452	356	324	2681	1505	927	596	418	2242	974	562	397	338
	.01	937	473	316	249	223	1906	1073	659	433	305	1595	687	392	282	234
	.05	666	326	218	164	144	1362	765	467	313	219	1117	484	275	191	152
10%	.0001	937	446	291	233	218	1430	763	467	313	227	1561	647	366	260	224
	.001	722	343	228	176	160	1133	596	370	244	176	1218	504	284	201	167
	.01	516	241	158	123	108	804	428	267	178	128	870	358	197	140	116
	.05	363	169	109	84	71	565	307	190	127	92	618	252	137	95	76
25%	.0001	556	202	121	94	85	568	247	145	100	75	900	296	155	103	88
	.001	434	155	96	72	63	450	194	115	78	59	710	233	119	81	67
	.01	308	110	66	49	43	318	139	82	57	42	504	165	84	54	46
	.05	214	76	45	33	27	229	101	60	39	30	354	116	57	39	28

An asymptotic approximation of the power of the TDT can be obtained in the following way:

(i)   Calculate the $(1 - \alpha)$-quantil $\chi^2_{1,1-\alpha}$ of the central χ^2 distribution with one degree of freedom.

(ii) Define:

$$R: = p^2\Psi_2 + 2pq\Psi_1 + q^2$$

$$a: = \frac{2pq}{R}(p\Psi_2 + (1 - 2p)\Psi_1 - q)$$

$$b: = \frac{2pq}{R^2}(R^2 + 2pq\Psi_2 + q^2\Psi_1 + p^2\Psi_2\Psi_1)$$

$$c: = \frac{2pq}{R}(p\Psi_2 + \Psi_1 + q)$$

(Eq. 12.10)

and calculate:

$$x: = \frac{c}{b} * \chi^2_{1,1-\alpha}$$

$$\lambda: = n * \frac{a^2}{b}$$

(Eq. 12.11)

(iii) The approximate power of the TDT for sample size n and Ψ_2, Ψ_1, p equals the probability that a χ^2 distribution with noncentrality parameter λ is larger than x, that is:

$$\text{power} \approx P(\chi^2_{1,\lambda} > x)$$

(Eq. 12.12)

The advantage of this approximative power calculation is that it can be performed much faster than simulations. Its only requirement is a program that calculates the distribution function of the noncentral χ^2 distribution as, for example, the PROBCHI-function of the SAS-system (SAS Institute Inc., 1990). Its disadvantage is that its result is not always very precise, but it can still be used as a starting value for the simulations.

If assumption (A1) is not satisfied so that there is positive distance between marker and disease locus (i.e. $\Theta > 0$), then the sample size has to be increased by roughly the factor $1/(1 - 2\Theta)^2$ to obtain the same power as for $\Theta = 0$ (Knapp et al., 1993). For the power calculations by simulation, assumption (A2) is not crucial. Without (A2), however, a larger number of parameters have to be specified to describe the true state of nature. For example, if the population is a mixture of different subpopulations, the number and proportions of these subpopulations and risk parameters for each subpopulation will have to be given.

169

Extensions

Throughout this paper it has been assumed that there is a particular allele (denoted by M) at a marker locus for which its association with the disease is of interest. When testing for linkage disequilibrium between a polymorphic marker locus and a disease without such a prior hypothesis, each marker allele in turn may be considered against all other alleles at the marker locus. But performing many tests necessitates a P value adjustment such as the Bonferoni correction and this can become quite conservative. The alternative is to apply an overall test for linkage disequilibrium between a polymorphic marker locus and a disease. The approach of Sham and Curtis (1995), which extends the TDT to a polymorphic marker locus, is motivated by the form of the joint distribution of alleles transmitted and not transmitted by parents of an affected child and can be implemented by standard software for logistic regression. Terwilliger (1995) suggested a parametric likelihood approach which is especially powerful in cases where there is a single allele at the marker locus with an increased frequency in haplotypes carrying the disease-causing gene. The general merits of Terwilliger's test as well as the null distribution of its test statistic are currently a matter of dispute (Sham *et al.*, 1996; Terwilliger, 1996).

Conclusion

Association studies are able to detect disease loci that would be missed by linkage analysis, for example, when this disease locus explains only a small fraction of the disease (Greenberg, 1993). Linkage disequilibrium can be expected to extend only over short distances. Therefore, association studies do not seem to be suited to performing a genome scan. In case of a large number of potential candidate genes, as for psychiatric disorders, most positive findings will be false positives due to the low prior probability (Crowe, 1993).

The affected family-based control method for association studies discussed in the present paper solves the problem of selecting an appropriate control group, but other general difficulties (e.g. the problems defining the disease phenotype and the question of an appropriate correction for multiple testing when multiple markers are studied) are not addressed by this approach.

170

References

Brewerton, D.A., Hart, F.D., Nicholls, A., Caffrey, M., James, D.C., Sturrock, R.D. Ankylosing spondylitis and HL-A 27. (1973) *Lancet* **1 (809):** 904–907.

Crow, J.F., Kimura, M. An Introduction to Population Genetics Theory. (1970) Harper and Row, New York.

Crowe, R.R. Candidate genes in psychiatry: an epidemiological perspective. (1993) *Am. J. Hum. Genet.* **48 (2):** 74–77.

Falk, C.T., Rubinstein, P. Haplotype relative risks: an easy reliable way to construct a proper control sample for risk calculations. (1987) *Ann. Hum. Genet.* **51 (3):** 227–233.

Fleiss, J.L. *Statistical Methods for Rates and Proportions.* (1973) John Wiley and Sons, New York.

Greenberg, D.A. Linkage analysis of "necessary" disease loci versus "susceptibility" loci. (1993) *Am. J. Hum. Genet.* **52 (1):** 135–143.

Knapp, M., Seuchter, S.A., Baur, M.P. The haplotype-relative-risk (HRR) method for analysis of association in nuclear families. (1993) *Am. J. Hum. Genet.* **52 (6):** 1085–1093.

Knapp, M., Wassmer, G., Baur, M.P. The relative efficiency of the Hardy–Weinberg equilibrium-likelihood and the conditional on parental genotype-likelihood methods for candidate-gene association studies. (1995) *Am. J. Hum. Genet.* **57 (6):** 1476–1485.

Lander, E.S., Schork, N.J. Genetic dissection of complex traits. (1994) *Science* **265 (5181):** 2037–2048.

McNemar, Q. Note on sampling error of the differences between correlated proportions or percentages. (1947) *Psychometrika* **12 (153):** 157.

Mourant, A.E., Kopec, A.C., Domaniewska–Solczak, K. *Blood Groups and Diseases* (1978) Oxford University Press, Oxford, New York.

Ott, J. Statistical properties of the haplotype relative risk. (1989) *Genet. Epidemiol.* **6 (1):** 127–130.

Rubinstein, P., Walker, M., Krassner, J., Carrier, C., Carpenter, C., Dobersen, M.J., Notkins, A.L., Mark, E.M., Nechemias, C., Hausknecht, R.U., Ginsberg Fellner, F. HLA antigens and islet cell antibodies in gestational diabetes. (1981) *Hum. Immunol.* **3 (3):** 271–275.

SAS Institute Inc. (1990) SAS® Language: Reference, Version 6. SAS Institute Inc., Cary.

Schaid, D.J., Sommer, S.S. Genotype relative risks: methods for design and analysis of candidate-gene association studies. (1993) *Am. J. Hum. Genet.* **53 (5):** 1114–1126.

Schaid, D.J., Sommer, S.S. Comparison of statistics for candidate-gene association studies using cases and parents. (1994) *Am. J. Hum. Genet.* **55 (2):** 402–409.

Sham, P.C., Curtis, D. An extended transmission/disequilibrium test (TDT) for multi-allele marker loci. (1995) *Ann. Hum. Genet.* **59 (3):** 323–336.

Sham, P.C., Curtis, D., MacLean, C.J. Likelihood ratio tests for linkage and linkage disequilibrium: asymptotic distribution and power. (1996) *Am. J. Hum. Genet.* **58 (5):** 1093–1095.

Sobell, J.L., Heston, L.L., Sommer, S.S. Delineation of genetic predisposition to multifactorial disease: a general approach on the threshold of feasibility. (1992) *Genomics* **12 (1):** 1–6.

Spielman, R.S., McGinnis, R.E., Ewens, W.J. Transmission test for linkage disequilibrium: the insulin gene region and insulin-dependent diabetes mellitus (IDDM). (1993) *Am. J. Hum. Genet.* **52 (3):** 506–516.

Terwilliger, J.D., Ott, J. A haplotype-based 'haplotype relative risk' approach to detecting allelic associations. (1992) *Hum. Hered.* **42 (6):** 337–346.

Terwilliger, J.D. A powerful likelihood method for the analysis of linkage disequilibrium between trait loci and one or more polymorphic marker loci. (1995) *Am. J. Hum. Genet.* **56 (3):** 777–787.

Terwilliger, J.D. Likelihood ratio tests for linkage and linkage disequilibrium: asymptotic distribution and power: reply. (1996) *Am. J. Hum. Genet.* **58 (5):** 1095–1096.

Tiwari, J.L., Terasaki, P.I. *HLA and Disease Association.* (1985) Springer, New York.

Chapter 13
A note on the relative power of association studies and linkage analysis in the genetic analysis of disease

Bertram Müller-Myhsok

Introduction

When studying a certain region or gene for involvement in a given disease by the methods of genetic mapping, two main strands of analysis are being commonly used. One is the analysis of the phenomenon of allelic association, the other the analysis of cosegregation (i.e. the analysis of linkage). Both phenomena may be seen as being closely related and it may be asked whether the exploitation of both phenomena may not greatly increase the power of a given study, especially in the context of the investigation of a candidate gene or candidate region for a disease.

We investigated whether there is indeed something to be gained from using both approaches jointly.

Material and methods

Family structure used

Figure 13.1 shows the pedigree structure used. It is a nuclear family with at least one child affected with the disease studied, another child and the parents.

GENETIC MAPPING OF DISEASE GENES
ISBN 0-12-232735-7

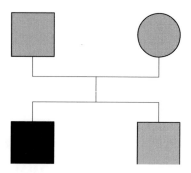

Figure 13.1 Pedigree structure used

The phenotype of the parents and the second child was not fixed to either affected or unaffected, but was left free to vary according to the genetic model studied.

Models considered

Type of model The models considered were diallelic disease models, that is we supposed two alleles at the disease locus ('the candidate gene'), which we termed A and a. Penetrances were modelled such that $f(aa) \geq f(Aa) \geq F(AA)$, with $f(AA)$ being set to 0. Therefore the presence of at least one copy of allele a was necessary for the development of the disease, no phenocopies being allowed for. The marker locus was also considered to be diallelic, with alleles B and b. Allelic association between marker alleles and the alleles at the disease locus was modelled via the coupling matrix, which is a matrix containing the probabilities to find a certain allele at the disease locus in the presence of a certain allele at the marker locus. This parametrisation is the one also used in the marker association segregation chi-square (MASC) method (Clerget-Darpoux *et al.*, 1988). All calculations were done based on the MASC-Kernel.

Constraints and space of models studied We put a risk constraint on the models studied (i.e. we only considered those models that gave a risk for the sibs of affected children of 5%). Recombination between the marker locus and the disease locus was neglected as we operated in a candidate gene setting. We searched the space of models within the framework mentioned in the previous section 'Type of model' in steps of 0.2 for the allele frequencies both at the marker locus and the disease locus and applied a finer grid of 0.01 to the penetrances possible. The values of the coupling matrix were allowed to vary freely in order to comply with the risk constraint. As a whole we found 100 models within this grid to fulfil the risk constraint.

Tests investigated

We investigated three tests: one test of association (antigen frequencies in affected persons, AGFAP), one of cosegregation (XYZ) and one using jointly the information of both the AGFAP and the XYZ test.

The AGFAP test The AGFAP test was described by Thomson (1983). It tests the antigen frequencies in affected persons by means of a χ^2 test of goodness-of-fit against the expected frequencies under a given model. It can, of course be used to test for association by using as the model tested a model that is in accordance with the null hypothesis of no association. It is a population-based and not a family-based test of association.

The XYZ test The XYZ test (Motro and Thomson, 1985) uses the information about whether sibs are identical (X), semi-identical (Y) or non-identical by descent (Z) at a marker locus to compare the proportions of X, Y and Z found against those expected under a given genetic model. If restricted to affected sibs and with the model used to test against being chosen as one that implies no linkage between the marker locus and the disease locus, it can be used as a regular affected sib pair test.

The combined approach The combined approach uses the information gained from both the AGFAP and the XYZ test to form a joint statistic by means of the simple transformation given by Fisher (1958), which allows the conversion of the significance value p found in one test into the value of a χ^2 variable with two degrees of freedom:

$$-2 \cdot ln\, p = \chi_2^2 \qquad \textbf{(Eq. 13.1)}$$

The p values obtained both from the independent XYZ and AGFAP tests may then be combined by simply adding the χ^2 values obtained following **Eq. 13.1** and the result may be compared to a χ^2 distribution with four degrees of freedom.

Comparing the performances of the tests For both the XYZ and the AGFAP test, as well as the combined approach the expected sample size to reject the respective null hypotheses (no linkage, no association, neither association nor linkage) at a significance level of 5% was calculated. The smallest value for a given underlying genetic model was termed the smallest expected sample size for that model. We then measured the difference to the smallest expected sample size for each of the tests considered. For the test that gave the smallest expected sample size this difference was zero, for the other tests it was typically greater than zero.

Results

A graphical representation of the distribution of the smallest expected sample sizes among the three tests considered is shown in *Figure 13.2*. In more than 50% of the models considered, the combined approach was the method of choice (i.e. it gave the smallest exected sample size in more than 50% of the cases).

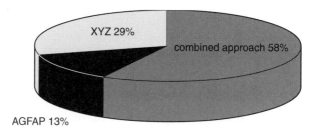

Figure 13.2 Percentages of smallest expected sample sizes

Table 13.1 gives another viewpoint on the result. As can be seen, the combined approach was at the maximum 49 observational units (i.e. nuclear families) away from the smallest expected sample size. For the XYZ test this figure was more than three times as high, while for the AGFAP test there were cases where there was no power at all (i.e. models in which there was no association between marker and disease alleles).

When we allowed for phenocopies (i.e. did not enforce the constraint $f(AA) = 0$ while otherwise retaining the other constraints on the parameters of the

Table 13.1. Deviations from the smallest expected sample size

Test	Maximum	Median	Mean
AGFAP	Infinity	86	Not meaningful
XYZ	181	20	35
Combined approach	49	0	7

models) the results were similar (*Figure 13.3*). The combined approach was still the approach of choice. The XYZ approach gave the smallest expected sample size in slightly more cases (38 models versus 37 models) than the joint approach, but the maximum, mean and median deviations (*Table 13.2*) spoke in favour of the joint approach.

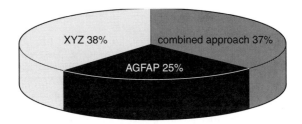

Figure 13.3 Percentages of smallest expected sample size allowing for phenocopies

Table 13.2. Deviations from the smallest expected sample size allowing for phenocopies

Test	Maximum	Median	Mean
AGFAP	Infinity	134	Not meaningful
XYZ	6327	82	565
Combined approach	605	6	75

Discussion

In complex genetic traits where the analysis of the role of candidate genes plays an important role, there is no single best approach for every possible genetic model, as shown above. However, there appears to be a safe strategy, which even when not the best method for the underlying (and unknown) genetic model consists of an approach combining association and linkage analysis. Such an approach tends to be conservative on the resources needed.

Spielman *et al.*, 1993 have also shown that a joint approach could be feasible for the transmission disequilibrium test (TDT) test, which does not suffer the population-based problems of the more conventional association tests such as the AGFAP. The power of a combined approach may be further enhanced by intelligent categorisation of the data, as implemented in the MASC method (Clerget-Darpoux *et al.*, 1988).

Acknowledgements

Thanks go to Françoise Clerget-Darpoux, Marie-Claude Babron and Maria Martinez for helpful discussion, encouragement and advice.

References

Clerget-Darpoux, F., Babron, M.C., Prum, B., Lathrop, G.M., Deschamps, I., Hors, J. A new method to test genetic models in HLA associated diseases: the MASC method. (1988) *Ann. Hum. Genet.* **52 (3):** 247–258.

Fisher, R.A. *Statistical Methods for Research Workers.* 6th edition. (1958) Oliver and Boyd, Edinburgh.

Motro, U., Thomson, G. The affected sib method. I. Statistical features of the affected sib-pair method. (1985) *Genetics* **110 (3):** 525–538.

Spielman, R.S., McGinnis, R.E., Ewens, W.J. Transmission test for linkage disequilibrium: the insulin gene region and insulin-dependent diabetes mellitus (IDDM). (1993) *Am. J. Hum. Genet.* **52 (3):** 506–516.

Thomson, G. Investigation of the mode of inheritance of the HLA associated diseases by the method of antigen genotype frequencies among diseased individuals. (1983) *Tiss. Ant.* **21 (2):** 81–104.

Chapter 14
Use of association and linkage information for the study of multifactorial diseases

Françoise Clerget-Darpoux

The etiology of many human diseases is complex and often involves a combination of genetic and environmental risk factors. Genetic risk factors can be common polymorphisms that are neither necessary nor sufficient to develop the disease. In such a case, the power to detect them may be poor.

Candidate genes that have a function potentially related to the disease represent possible risk factors for the disease. The investigation of a candidate gene usually includes two steps:

(1) the demonstration of its involvement as a risk factor for the disease;
(2) the modelling of its role.

The genotypes of the candidate gene are not always observable. The information on a genetic marker situated at or near the candidate gene locus can then be used.

Detection of risk factors

Two types of information may be provided by the marker on the presence of a risk factor. First, if there is a gametic disequilibrium between the marker alleles and the risk factor alleles, association can be observed between the marker and the disease at the population level. Certain marker alleles are more often present in affected individuals than in random individuals. The power to detect an

GENETIC MAPPING OF DISEASE GENES
ISBN 0-12-232735-7

association depends not only upon its strength, but also upon the characteristics of the marker. A marker with two (or not too many) alleles should be preferred. The marker must be close enough to the candidate gene to hope for gametic disequilibrium. In that respect the 'candidate region' where the markers are chosen has to be very small. Indeed in a large population panmictic for a long time, gametic disequilibrium generally exists only between alleles of very closely linked loci. However, one must be aware that association may be observed in the absence of linkage when cases and controls are not issued from the same homogeneous and panmictic population.

Second, if the marker and the disease are linked (dependent segregation in families), then relatives with the same or different disease status are more or less similar for the marker than expected under independence. In particular, under independent segregation of the disease and a marker, the sib pairs are expected to share 2, 1 and 0 alleles identical by descent (IBD) in the proportion 1/4, 1/2, 1/4, respectively. A deviation from this expectation may be tested by a conformity χ^2 test (IBD test) or using a maximum likelihood ratio (MLS)-test as proposed by Risch (1990). The MLS-test has the advantage to infer the IBD probabilities in ambiguous situations such as parent homozygote or untyped for the marker. It is sometimes referred to as a maximum logarithm of odds (lod) score. Note that this terminology is misleading since the MLS-test is not equivalent to the usual lod score test (Morton, 1955). Indeed, the parameters of interest are not the same. It is the IBD distribution in one case and the recombination fraction between the marker and disease loci in the other. The statistical properties are of course also different. Those of the MLS have been studied by Holmans (1993). The power to detect a linkage depends not only upon the characteristics of the risk factor, but also upon the informativity of the marker. A very polymorphic marker should be preferred. The polymorphism may be increased by haplotyping very closely linked markers.

Simultaneous use of gametic disequilibrium and linkage information: the transmission/disequilibrium test

Most methods for the investigation of the role of a candidate gene use only one type of information. The transmission/disequilibrium test (TDT) (Spielman *et al.*, 1993, for a marker with two alleles) incorporates both types of information. It compares the alleles transmitted by heterozygous parents to their affected child with the nontransmitted alleles and permits simultaneous testing for linkage and association. It has been extended to the situation of multiallele marker loci (Sham and Curtis, 1995; Bickeböller and Clerget-Darpoux, 1995).

The alleles not transmitted to the affected child(ren) create an internal family-based control group permitting estimation of relative risk (Falk and Rubinstein, 1987; Terwilliger and Ott, 1992; Knapp *et al.*, 1993). The non-transmitted allele sample also provides, as shown by Thomson (1995), unbiased allele frequency estimates.

Power comparison of the TDT versus the IBD test to detect linkage

The power of the TDT relative to that of the IBD test depends on both the underlying genetic model and on the available family data. Families with two affected sibs are always more informative than those with one affected and one unaffected child. The IBD test is always more powerful in the first situation and, contrary to the TDT, is independent of the presence of gametic disequilibrium. When there is strong linkage disequilibrium, the TDT can be more powerful than the IBD test. In that case, linkage can be detected by the TDT, even in families with only one affected child (Clerget-Darpoux *et al.*, 1995).

The IBD test taking into account gametic disequilibrium

When there is gametic disequilibrium between the marker and risk factor alleles, then the expectation of IBD sharing for an affected sib of an index patient depends upon the marker genotype of this index (Clerget-Darpoux *et al.*, 1995). Like the TDT, it is possible to test simultaneously for linkage and association in testing the homogeneity of the IBD kl distribution in sib pairs conditioned on the genotype formed with marker allele k and marker allele l of the index case (stratified IBD distribution).

For a marker locus with m alleles, the homogeneity of the IBD kl distributions can be tested by a χ^2 with $m^2 + m - 2$ degrees of freedom. Note that, under a recessive disease model, homogeneity of the IBD_{kl} distributions is also expected. Consequently, gametic disequilibrium cannot be detected in this situation.

Modelling the role of a risk factor

The two types of information (gametic disequilibrium and linkage) can also be used to model the role of a risk factor.

The information provided by gametic disequilibrium between the marker alleles and the risk factor alleles can be used through the antigen genotype frequencies among patients (AGFAP)-method proposed by Thomson (1983) and extended by Risch (1983). A genetic model for the risk factor is tested by comparing the marker genotype expected under this model to the one observed in a sample of unrelated patients.

The information provided by linkage between the marker and the risk factor locus can be used through the IBD method proposed by Thomson and Bodmer (1977). A genetic model for the risk factor is tested by comparing the IBD distribution expected under this model to the one observed in a sample of affected sib pairs.

If there is gametic disequilibrium, the power of model discrimination may be increased by stratifying the affected sib pair sample according to the index marker genotypes and using the stratified IBD distribution.

The marker association segregation chi-square method

The marker association segregation chi-square (MASC)-method (Clerget-Darpoux *et al.*, 1988) is designed to model the role of a candidate gene in a complex disease using both linkage and association information. The analysis is performed on nuclear families of index patients. The index patients are classified into nested categories in three steps according to:

(1)   the clinical statuses of their parents and sibs (familial configuration);
(2)   their marker genotype;
(3)   their IBD status with a sib.

The MASC method is a generalization of the AGFAP method and of the stratified affected sib pair method. The index patients can be categorized further to incorporate more information or to test more complex models such as the role of two candidate genes and their interactions (Dizier *et al.*, 1994).

The method is based on the comparison of the observed categorized data to the expected data under various hypotheses, in particular under the hypothesis of no involvement of the candidate gene.

The categories formed in each step are nested in the categories of the previous step. Consequently the outcomes are stochastically independent in each category of the step *i* and are also independent of the outcomes of the other steps. In addition to the property of independence, the choice of the categories is made in order to use the most relevant information. In the case of unspecified mode of ascertainment for the index cases, only steps 2 and 3 are used, but conditionally to the step 1 categorization.

For a given genetic model, the expected number in each category of step i is computed conditionally on the observed number in the corresponding category of step $i - 1$. For each classification, the fit of genetic model to observed data is tested by a χ^2 statistic. Because of the independence of all these statistics, their sum Θ follows a χ^2 with the sum of the degrees of freedom of each statistic (Prum *et al.*, 1990). The program MASC computes the expected values in each category and then the Θ value and minimizes it in the parameter space.

Some categories may be discarded or pooled with others so that the expected number in each category is large enough to apply a χ^2 test. However, this means a loss of the information contained in the data and may greatly decrease the power of discrimination between models. A test based on the exact probability distribution of a test statistic has been implemented to avoid this dilemma (Müller and Clerget-Darpoux, 1991).

The MASC method has been applied to model the role of the human leukocyte antigen (HLA) complex for different diseases of the immune system: insulin-dependent diabetes mellitus (IDDM) (Clerget-Darpoux *et al.*, 1991), rheumatoid arthritis (Dizier *et al.*, 1993), multiple sclerosis (Clerget-Darpoux and Babron, 1993) and celiac disease (Clerget-Darpoux *et al.*, 1994). The gain of information on the modelling of a risk factor by using the MASC method is illustrated below on IDDM data analysis.

Illustration on IDDM data

Risk factor in the HLA region

In Insulin Dependent Diabetes (IDDM), association between some HLA antigens (in particular DR3 and DR4 in Caucasian populations) and the disease has been shown. Distortion in the segregation of the HLA markers in families with affected individuals have also been reported, unambiguously showing the existence of risk factor(s) in the HLA region.

In a sample of 158 Caucasian IDDM sib pairs, the proportion of those sharing two, one or zero haplotypes are 0.56, 0.38 and 0.06, respectively. This IBD distribution fits the effect of a recessive risk factor allele with a frequency of 0.35 very well. However, such a model does not explain the AGFAP distribution at the DR locus, in particular, the high proportion of heterozygotes DR3DR4 compared to the proportion of homozygotes DR3DR3 or DR4DR4. Furthermore, it does not explain the difference in the IBD distributions when stratified on the DR genotype index cases. When the index cases are DR3DR4, the proportion of sibs sharing two HLA haplotypes is 71%, whereas it is only 46% when the index cases are DR4 non DR3. It is obvious that the contradiction in the model fitting is only due to the fact that these approaches take into account different information and that the maximum information must be used simultaneously. We analysed 416 HLA-typed affected Caucasian individuals and their relatives by the MASC method (Clerget-Darpoux *et al.*, 1991) in order to explain the whole observation.

We considered the 'classical' models with one or two susceptibility alleles at one locus, and because of its biological interest we considered the possibility of a susceptibility molecule formed by the complementation of two specific alleles at two loci of the HLA region. In this spirit, Khalil *et al.* (1990) proposed a suscepti-bility molecule composed of a DQα chain bearing an arginine at position 52, arg52$^{(+)}$, and a DQβ chain lacking an aspartic acid at position 57, asp57$^{(-)}$. This particular complementation model was also tested at the molecular level.

Using MASC, we demonstrated that among the models we considered, only the complementation of two alleles at two distinct but strictly linked loci was able to explain our data. In searching for such a heterodimer, it appears, in view of our results, that molecular reseachers should not look for a sequence shared by both DR3 and DR4 haplotypes. On the contrary, they should focus on finding a com-bination of two sequences that differentiate these haplotypes. In particular, we showed that the complementation of asp57$^{(-)}$ and arg52$^{(+)}$ cannot explain the HLA observations. MASC may provide an interactive link between the molecular approach and the genetic modelling in the study of IDDM. The HLA component of the genetic susceptibility to IDDM is clearly complex.

Risk factor in the insulin region

Other familial factors, either genetic or environmental, are also involved. In par-ticular, the role of a risk factor in the insulin gene region has been clearly demon-strated. The demonstration of this risk factor provides a good illustration of how information can be obtained from gametic disequilibrium.

Bell *et al.* (1984) showed a strong association between one allele (C1) of 5′flanking polymorphism (5′FP) of the insulin gene and IDDM. This result moti-vated the investigation of this polymorphism on an affected IDDM sib pair sam-ple collected for a Genetic Analysis Workshop held in 1987. The IBD distribution for the insulin region markers obtained on 95 IDDM affected sib pairs (0.24, 0.52 and 0.24) did not show any distortion when compared to that expected under inde-pendent segregation of the insulin gene and IDDM. However, a clear difference between the parental haplotypes transmitted and unstransmitted to the diabetic members was shown by several teams (Field, 1989; Thomson *et al.*, 1989). More recently, as an illustration of the TDT, Spielman *et al.* (1993) showed that in this same family sample, the 57 parents heterozygous for the class 1 allele of 5'FP transmitted the class 1 allele to affected individuals 78 times and did not transmit it 46 times. As demonstrated in the same paper, this strong difference permits the conclusion of simultaneous linkage and gametic disequilibrium ($p = 0.004$). This illustrates that in some situations the TDT may be more powerful than IBD test for detecting linkage.

The same evidence can be obtained from considering the IBD distribution stratified on index genotypes for 5'FP. It is interesting to note that the proportion of sibs sharing two IBD insulin haplotypes with the index cases is greater than 1/4 when the index is C1C1, but lower than 1/4 in the other situations (*Table 14.1*).

Table 14.1. Identity by descent sharing distribution for the insulin gene: in the whole sample of affected sib pairs; in the subsample where the index is homozygous for the C1 allele of 5FP; in the remaining sample of affected sib pairs (index different from C1C1)

	Observed	Expected under M0
Global IBD	(0.24,0.52,0.24)	(0.26,0.50,0.24)
IBD for sibs of C1C1 index	(0.28,0.55,0.16)	(0.29,0.50,0.21)
IBD for sibs of non-C1C1 index	(0.10,0.47,0.43)	(0.20,0.50,0.30)

Applying MASC to the observations on the insulin 5'FP and HLA indicates a simultaneous effect of risk factors in the HLA and insulin regions (Dizier, Babron and Clerget-Darpoux, 1994). A multiplicative effect of HLA and insulin was not rejected and the best fit model M0 includes a complementation of susceptibility alleles in the HLA region and a complete linkage disequilibrium between the insulin region risk factor and the class 1 allele of 5'FP. This model explains the stratification observed in the IBD distribution for the insulin region (see the expected distributions in *Table 14.1*). Note that, when there is gametic disequilibrium, decreased IBD sharing of affected sib pairs in a subgroup may be observed. The complete gametic disequilibrium between C1 and the insulin risk factor is in favor of a direct implication of 5'FP. This has been also suggested by Bennet *et al.* (1995).

IDDM risk factors in other parts of the genome have been shown recently by linkage analyses in a systematic screening approach (Todd, 1995). It has revealed some regions of interest in the genome, but fine mapping of such regions is still a difficult challenge with the current state of art. Finding markers in gametic disequilibrium with the risk factor would permit further modelling of risk factors and unravelling of the genetic determinism of IDDM.

The MASC method can be applied to the study of any complex disease provided there is information on the polymorphic markers of candidate genes. The greater the marker polymorphism and the disequilibrium between marker alleles and the candidate gene alleles, the better the information on the potential role of the candidate gene. The advantage of the MASC method is to potentiate the information provided by linkage and gametic disequilibrium.

References

Bell, G.I., Horita, S., Karam, J.H. A polymorphic locus near the human insulin gene is associated with insulin-dependent diabetes mellitus. (1984) *Diabetes* **33** (2): 176–183.

Bennett, S.T., Lucassen, A.M., Gough, S.C., Powell, E.E., Undlien, D.E., Pritchard, L.E., Merriman, M.E., Kawaguchi, Y., Dronsfield, M.J., Pociot, F., *et al.* Susceptibility to human type 1 diabetes at IDDM2 is determined by tandem repeat variation at the insulin gene minisatellite locus. (1995) *Nat. Genet.* **9** (3): 284–292.

Bickeböller, H., Clerget-Darpoux, F. Statistical properties of the allelic and genotypic transmission/disequilibrium test for multiallelic markers. (1995) *Genet. Epidemiol.* **12** (6): 865–870.

Clerget-Darpoux, F., Babron, M.C., Prum, B., Lathrop, G.M., Deschamps, I., Hors, J. A new method to test genetic models in HLA associated diseases: the MASC method. (1988) *Ann. Hum. Genet.* **52** (3): 247–258.

Clerget-Darpoux, F., Babron, M.C., Deschamps, I., Hors, J. Complementation and maternal effect in insulin-dependent diabetes. (1991) *Am. J. Hum. Genet.* **49** (1): 42–48.

Clerget-Darpoux, F., Babron, M.C. HLA-Sclérose en plaque inférences génétiques. Dans: A. Alpéroritch, J. Hors, O. Lyon-Caen (eds). *Jumeaux et Sclérose en Plaque.* Paris, John Libbey Eurotext, 1993, pp. 5–10.

Clerget-Darpoux, F., Bouguerra, F., Kastally, R., Semana, G., Babron, M.C., Debbabi, A., Bennaceur, B., Eliaou, J.F. High risk genotypes for celiac disease. (1994) *C. R. Acad. Sci. III.* **317** (10): 931–936.

Clerget-Darpoux, F., Babron, M.C., Bickeböller, H. Comparing the power of linkage detection by the transmission disequilibrium test and the identity by descent test. (1995) *Genet. Epidemiol.* **12** (6): 583–588.

Dizier, M.H., Babron, M.C., Clerget-Darpoux, F. Interactive effect of two candidate genes in a disease: extension of the marker-association-segregation chi-square method. (1994) *Am. J. Hum. Genet.* **55** (5): 1042–1049.

Dizier, M.H., Eliaou, J.F., Babron, M.C., Combe, B., Sany, J., Clot, J., Clerget-Darpoux, F. Investigation of the HLA component involved in rheumatoid arthritis (RA) by using the marker association-segregation chi-square (MASC) method: rejection of the unifying-shared-epitope hypothesis. (1993) *Am. J. Hum. Genet.* **53** (3): 715–721.

Falk, C.T., Rubinstein, P. Haplotype relative risks: an easy reliable way to construct a proper control sample for risk calculations. (1987) *Ann. Hum. Genet.* **51** (3): 227–233.

Field, L.L. Genes predisposing to IDDM in multiplex families. (1989) *Genet. Epidemiol.* **6** (1): 101–106.

Holmans, P. Asymptotic properties of affected-sib-pair linkage analysis. (1993) *Am. J. Hum. Genet.* **52** (2): 362–374.

Khalil, I., d'Auriol, L., Gobet, M., Morin, L., Lepage, V., Deschamps, I., Park, M.S., Degos, L., Galibert, F., Hors, J. A combination of HLA-DQ beta Asp57-negative and HLA DQ alpha Arg52 confers susceptibility to insulin-dependent diabetes mellitus. (1990) *J. Clin. Invest.* **85 (4):** 1315–1319.

Knapp, M., Seuchter, S.A., Baur, M.P. The haplotype-relative-risk (HRR) method for analysis of association in nuclear families. (1993) *Am. J. Hum. Genet.* **52 (6):** 1085–1093.

Morton, N.E. Sequential tests for the detection of linkage. (1955) *Am. J. Hum. Genet.* **7:** 277–318.

Müller, B., Clerget-Darpoux, F. A test based on the exact probability distribution of the chi² statistic – incorporation into the MASC method. (1991) *Ann. Hum. Genet.* **55 (1):** 69–75.

Prum, B., Guilloud Bataille, M., Clerget-Darpoux, F. On the use of chi-square tests for nested categorized data. (1990) *Ann. Hum. Genet.* **54 (4):** 315–320.

Risch, N. A general model for disease-marker association. (1983) *Ann. Hum. Genet.* **47 (3):** 245–252.

Risch, N. Linkage strategies for genetically complex traits. II. The power of affected relative pairs. (1990) *Am. J. Hum. Genet.* **46 (2):** 229–241.

Sham, P.C., Curtis, D. An extended transmission/disequilibrium test (TDT) for multi-allele marker loci. (1995) *Ann. Hum. Genet.* **59 (3):** 323–336.

Spielman, R.S., McGinnis, R.E., Ewens, W.J. Transmission test for linkage disequilibrium: the insulin gene region and insulin-dependent diabetes mellitus (IDDM). (1993) *Am. J. Hum. Genet.* **52 (3):** 506–516.

Terwilliger, J.D., Ott, J. A haplotype-based 'haplotype relative risk' approach to detecting allelic associations. (1992) *Hum. Hered.* **42 (6):** 337–346.

Thomson, G. Investigation of the mode of inheritance of the HLA associated diseases by the method of antigen genotype frequencies among diseased individuals. (1983) *Tissue Antigens* **21 (2):** 81–104.

Thomson, G., Bodmer, W.F. The genetic analysis of HLA and disease association (1977). In Dausset, J., Svejgaard, A. (eds). *HLA and Disease*, pp. 84–93. Munksgaard, Copenhagen.

Thomson, G., Robinson, W.P., Kuhner, M.K., Joe, S., Klitz, W. HLA and insulin gene associations with IDDM. (1989) *Genet. Epidemiol.* **6 (1):** 155–160.

Thomson, G. Mapping disease genes: family-based association studies (1995) *Am. J. Hum. Genet.* **57 (2):** 487–498.

Todd, J.A. Genetic analysis of type 1 diabetes using whole genome approaches. (1995) *Proc. Nat. Acad. Sci. USA* **92 (19):** 8560–8565.

Chapter 15
SimIBD: a powerful robust nonparametric method for detecting linkage in general pedigrees

Sean Davis, Lynn R. Goldin and Daniel E. Weeks

Summary

Nonparametric methods that do not rely on a strict affected sib pair (ASP) study design, but allow for any family structure provide powerful tools for mapping genes underlying complex traits. The affected pedigree member (APM) method is useful for analysing such extended families, but has several shortcomings. We present here simulation-based statistics that overcome several of the disadvantages of the APM-method and provide a powerful test for linkage. Our simulation-based statistics, SimAPM and SimIBD, use a conditional simulation approach to produce an empirical null distribution and empirical p value. SimAPM ignores genotypes in unaffected individuals and measures identity by state (IBS) sharing. SimIBD uses available genotypes in the unaffected individuals to measure identity by descent (IBD) sharing. SimAPM and SimIBD are essentially the same when there are no typed unaffected individuals. Our statistics are robust to allele frequency misspecification. We compare our statistics to other nonparametric methods that can handle families larger than nuclear families and find that when determining the statistical significance of the observed data, our simulation-based statistics can be more powerful than similar statistics that use large-sample theory.

GENETIC MAPPING OF DISEASE GENES
ISBN 0-12-232735-7

Introduction

The power of traditional logarithm of odds (lod) score techniques to map genes responsible for inherited disorders has led to the discovery of numerous disease genes. These parametric methods, as they are often called, require that the mode of inheritance (in terms of the penetrance and frequency of the disease gene) is specified. However, faced with the prospect of mapping increasingly complex inherited disorders in which the mode of inheritance is unclear, many researchers have turned to other methods that do not rely on specifying a mode of inheritance for the disorder. Called nonparametric methods, this class of tests for linkage is especially useful for detecting genes associated with these complex disorders (Lander and Schork, 1994).

Most nonparametric statistics assay sharing of alleles at a marker locus. Allele sharing is often measured in terms of IBS sharing (i.e. two alleles are IBS if they have the same label regardless of ancestral origin) or IBD sharing (i.e. two alleles are IBD if they are IBS and of the same ancestral origin). When allele sharing, as measured by IBD or IBS sharing, is increased significantly from what is expected under the hypothesis of no linkage, statistical evidence exists for linkage between the marker and the disease.

The ASP-methods are designed to analyze data in nuclear families and look for increased allele sharing between ASPs. Methods that measure IBD sharing are generally more powerful than those that measure IBS sharing because IBS status does not uniquely determine IBD status. However, determining IBD unambiguously in the face of incompletely typed or partially informative data is often impossible. Therefore, likelihood-ratio tests have been developed (e.g. Risch, 1990; Kruglyak et al., 1996) that use all the available information to assess IBD sharing and are therefore robust to missing information. The ASP paradigm is becoming an increasingly common study design for searching for disease-susceptibility loci. However, the ASP-methods are, by definition, applicable only to nuclear family data, so analyzing larger pedigrees generally requires breaking up the large family and then treating them (incorrectly) as independent nuclear families; alternatively, large families can be excluded from ASP analyses.

To treat pedigrees with extended family structures correctly, the APM-method (Weeks and Lange, 1988; Weeks and Lange, 1992) was developed. The APM-method tests whether IBS sharing between all affected relative pairs is significantly increased compared to that expected under the null hypothesis. Therefore, in the analysis of extended pedigrees, APM has an advantage over the ASP-methods because it uses information obtained from the entire pedigree (including the pedigree structure itself) rather than limiting analyses to the constituent nuclear families. The original APM-method (Weeks and Lange, 1988;

190

Weeks and Lange, 1992) has several weaknesses. First, the actual null distribution of the APM-method may be slightly skewed while the asymptotic null distribution is normally distributed; this can result in an elevated false-positive rate (Weeks and Lange, 1988; Weeks and Lange, 1992). It is now recommended that simulation is used to determine empirical p values. Second, the APM-method is sensitive to misspecification of the marker allele frequencies. Third, because the APM-method uses IBS status instead of IBD sharing, it ignores any genotype information in unaffected individuals and has low power compared to other linkage methods that do use IBD rather than IBS (Davis *et al.*, 1996).

To address the first two shortcomings of the APM-method, we developed SimAPM. SimAPM uses simulation of marker genotypes in the affected individuals conditional on the marker genotypes in the unaffected individuals to determine a conditional empirical null distribution. A slightly modified version of FASTSLINK (Ott 1989; Weeks *et al.*, 1990; Cottingham *et al.*, 1993), a general purpose simulation program, performs the simulations. A p value from SimAPM is based on the conditional empirical null distribution and is not plagued by the elevated false-positive rate seen with the p value based on the asymptotically normal statistic. Also, because the simulation of the affected individuals' marker genotypes is conditional on the marker genotypes of the unaffected individuals, sensitivity to allele frequency misspecification can be greatly reduced if marker genotypes are available for one or more unaffected people or eliminated if all founders are unaffected and typed. However, SimAPM, like APM, still measures IBS rather than IBD information, so it also has relatively low power (Davis *et al.*, 1996).

To increase power, thereby overcoming the final shortcoming of the APM-method, we developed SimIBD, which uses the same conditional simulation techniques as SimAPM, to determine a conditional empirical p value. However, SimIBD measures IBD sharing (as the name implies) between all relative pairs rather than measuring IBS sharing. Using IBD status as the measure of marker similarity provides a more powerful test of linkage than using IBS status (Lander and Schork, 1994; Davis *et al.*, 1996). SimIBD is robust to allele frequency misspecification, produces a conditional empirical p value, and is much more powerful than the APM-method. In addition, it can deal with pedigrees of arbitrary size and structure (including loops).

In this paper, we briefly recount the methods used by SimIBD and SimAPM and provide an example application of the SimIBD recursion algorithm. We give false-positive rates for SimAPM and SimIBD and investigate the effects of allele frequency misspecification. Finally, we have analyzed several data sets generated under a two-locus disease model (Martinez and Goldin, 1990; Goldin and Weeks, 1993) and present the power to detect linkage on these data for SimAPM, SimIBD, and Genehunter (Kruglyak *et al.*, 1996), a recently released program that performs both parametric and nonparametric linkage analyses on small- to medium-sized families and is not restricted to nuclear families.

Methods

The methods summarized here have been previously presented in detail (Davis *et al.*, 1996).

The simulated statistic

SimAPM and SimIBD use the same general steps to produce a *p* value based on the empirical null distribution. The steps are:

(1) Calculate and store the sharing statistic calculated for the original data (the observed statistic).
(2) Simulate many replicates of the affected individuals' marker genotypes conditional on the marker genotypes in the unaffected individuals to construct the null distribution. Simulation is done at a single locus, ignoring disease status.
(3) Calculate a value for the sharing statistic for each replicate of simulated data, yielding a point in the null distribution for each replicate.
(4) Determine the empirical *p* value by determining the proportion of points in the null distribution that have a greater value of the sharing statistic than the observed statistic.

The simulation in step (2) is done using FASTSLINK (Ott, 1989; Weeks *et al.*, 1990; Cottingham *et al.*, 1993) to simulate affected individuals' genotypes conditional on other typing information contained in each pedigree. Our program modifies LINKAGE input files automatically to instruct FASTSLINK to conditionally simulate marker genotypes for only the affected individuals. The marker genotypes of the unaffected individuals are not altered. The simulations are relatively fast because we are simulating marker genotypes at a single locus (the marker locus) and not at two loci (the marker and disease).

Approximation of the IBD probabilities

Our simulation-based statistics use the following general approach. Assume a pedigree of n members whose marker genotypes are described by $G = \{i_1, i_2, n_1, n_2\}$

where i_1 and i_2 are arbitrarily ordered alleles of individual i. We denote two alleles i_x and j_y that are IBD by $i_x \equiv j_y$. Thus, we define $P(i_x \equiv j_y | G)$ as the probability that allele i_x is IBD to allele j_y given the marker genotypes. This general approach is as follows.

(1) Choose two affected individuals i and j, who are not a parent–child pair.

(2) Assign a value to α_{i_x, j_y} as a measure of $P(i_x \equiv j_y | G)$: (A) For the SimAPM statistic, $\alpha_{i_x, j_y} = 1.0$ when i_x and j_y are IBS and 0.0 when i_x and j_y are not IBS (*Table 15.1*), that is, the estimated $P(i_x \equiv j_y | G)$ is simply the probability that i_x and j_y are IBS; (B) For the SimIBD statistic, we set α_{i_x, j_y} to 1.0 if the alleles are IBD and to 0.0 if the alleles are not IBD (see *Table 15.1*); if the IBD status is ambiguous ('maybe IBD'), we set α_{i_x, j_y} to the value returned by the recursive algorithm described below.

Table 15.1. Values of the SimAPM and SimIBD statistics for one allele from each of two affected individuals. i_x and j_y are two alleles chosen from affected individuals i and j

Identity status of two alleles	SimAPM value	SimIBD value
(1) Not IBS, not IBD	0	0
(2) IBS, not IBD	1	0
(3) IBS, maybe IBD	1	$\alpha = P(i_x \equiv j_y | G)*$
(4) IBS and IBD	1	1

*The SimIBD value is only an approximation when incomplete marker typing exists.

(3) Compute Z_{ij}, the pairwise statistic for affected individuals i and j, by the following formula where p_a = population frequency of allele i_a and p_b = population frequency of allele i_b and $p_a = p_b$ (because i_a and i_b are IBS):

$$Z_{ij} = \sum_{a=1}^{2} \sum_{b=1}^{2} f(p_a) \alpha_{i_a j_b} \qquad \text{(Eq. 15.1)}$$

where $f(p_a)$ is given by one of three functions: $\dfrac{1}{p_a}, \dfrac{1}{\sqrt{p_a}}$, or 1, as in the original APM statistic (Weeks and Lange, 1988). We use the function

$f(p_a) = \dfrac{1}{\sqrt{p_a}}$ throughout this paper for reasons presented elsewhere (Weeks and Lange, 1988)

(4) Sum the pairwise statistics over all pairs of affecteds (i, j) to arrive at the value of the statistic for the pedigree p:

$$Z_p = \frac{1}{\sqrt{r-1}} \sum_{i=1}^{r} \sum_{j=1}^{r} Z_{ij} \qquad \text{(Eq. 15.2)}$$

where r is the number of genotyped affected individuals in the pedigree.

(5) Sum the pedigree statistics Z_p across the M pedigrees to obtain the overall similarity statistic Z for the whole data set:

$$Z = \sum_{p=1}^{M} Z_p \qquad \text{(Eq. 15.3)}$$

Finding $P(i_x \equiv j_y | G)$

We have developed a recursive algorithm that will find $P(i_x \equiv j_y | G)$ exactly if all ancestors of i and j are fully-typed, but approximates it otherwise (Davis *et al.*, 1996). The recursive algorithm correctly handles inbreeding loops. Our algorithm works by moving up through the pedigree, 'expanding' all the IBD sharing information of lower generations onto each higher generation until we have reached all founders. Then we combine the information that has been collapsed onto the founders to determine $P(i_x \equiv j_y | G)$. We present here an intuitive example that informally illustrates the main concepts underlying our algorithm.

For the following example, we refer to *Figure 15.1*. Individuals 5, 6, and 7 are affected with a disease. Unordered marker genotypes are given under each individual. Let us first examine the IBD sharing between individual 5 and individual 7. This pair is identically heterozygous, so they share two alleles IBS. We are interested in finding the IBD sharing. Looking first at the 2 allele from individual 5, we observe that the origin of this allele is the 2 allele in individual 2. Turning to the 2 allele from individual 7, we easily determine that this allele is inherited from individual 4 with 100% certainty. With this step, we have reduced the problem from finding the ancestral origin of the allele in individual 7 to that of finding the origin of the same allele (IBD) in individual 4 and have reduced the number of generations to the top founders by one. Individual 4 obviously inherits her 2 allele from individual 2 with 100% certainty. We have reduced the IBD sharing probabilities for both individual 5 and individual 7 onto individual 2, a founder. Therefore, the probability that both individual 7 and individual 5 inherited the allele 2 from individual 2 is simply P (individual 7 received allele 2 from individual 2) P (individual 5 received allele 2 from individual 2) = (1.0)(1.0) = 1.0 (see case 4 in *Table 15.1*). Using similar logic for allele 3 in the same two affected individuals (5 and 7), we see that the 3 allele from individual

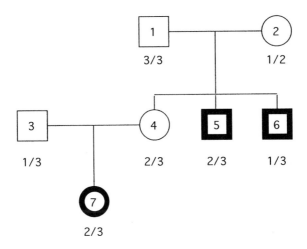

Figure 15.1 Example pedigree for computing IBD sharing probabilities

5 could have come from either of the alleles in individual 1, while the 3 allele from individual 7 had to come from individual 3. This is a case in which the 3 alleles are not IBD (see case 2 in *Table 15.1*), receiving a score of 0.0.

Sometimes, it is not possible to unambiguously determine IBD (see case 3 in *Table 15.1*). Individuals 5 and 6 both share a 3 allele that is inherited from individual 1. However, without phase information, IBD sharing cannot be determined absolutely. So, for individual 5, allele 3 is inherited with 50% probability from the left allele 3 in individual 1 and 50% probability from the right allele 3 in individual 1 and likewise for allele 3 in individual 6. Therefore, there are two possible ways that the 3 alleles from individuals 5 and 6 can be IBD. The probability that individuals 5 and 6 both inherited the left allele from individual 1 is 0.50 × 0.50 = 0.25. Likewise, the probability that both individuals inherited the right allele 3 from individual 1 is also 0.25. Since either the first case or the second case results in IBD sharing, the total probability of IBD for the 3 alleles is 0.25 + 0.25 = 0.5.

Untyped individuals in any ancestors of a pair of affected individuals introduce further uncertainty as to the pair's IBD status (beyond that already present due to any partial informativity such as homozygosity at the marker locus). In LOD score analyses, the traditional way of dealing with such uncertainty is to loop over all possible genotypes of the untyped individuals, adding in a partial statistic for each genotype. However, using such an approach resulted in a very slow program since these time-consuming summations had to be carried out for every replicate of simulated data. Instead, we use an approximation technique (see Davis *et al.*, 1996 for a detailed description). This approximation has the following property that when only the affected individuals are typed using SimIBD is identical to using SimAPM. Therefore, the power of SimAPM to detect linkage is a lower limit on the power of SimIBD that is realized when no unaffected pedigree members are typed.

Comparison of linkage methods

Data originally simulated by Goldin and Weeks (1993) using the method of Martinez and Goldin (1990) were analyzed in order to evaluate our statistics. Martinez and Goldin simulated their data assuming a two-locus disease model with a single marker (with four equally frequent alleles) linked at $\theta = 0.05$ to the first disease locus. Models involving two dominant loci (DD), two recessive loci (RR), and a dominant and a recessive locus (DR and RD) represent the epistatic models. A model with additive penetrance (AD) was also analyzed. The parameters used in the simulation of these five epistatic models correspond to those of unipolar and bipolar affective disorders and predict a population prevalence of 7% and a recurrence risk in first degree relatives of 25–30%. In addition, three models included heterogeneity at levels of 50% (H_{50}), 25% (H_{25}), and 10% (H_{10}), respectively, of families linked to the marker. The population prevalence for these models was 2% and disease could be due to either of the two dominant loci (90% penetrant). A simulated pedigree was selected to be in the data set if two or more affected individuals were present in each of the three sibships; the pedigree structure is shown in *Figure 15.2a*. For each model, the simulated data consisted of 50 replicates of 20 families for a total of 1,000 families per model and 8,000 families overall.

We chose to compare the power of our statistics to the power given by Genehunter (Kruglyak *et al.*, 1996). Genehunter, like APM, SimAPM, and SimIBD, uses a score function measuring allele sharing between affected individuals when performing nonparametric linkage analysis. Two options for score functions are offered. The first score function, which we will call Genehunter-Pairs, measures all pairwise allele IBD sharing between affected pairs. The second score function, here called Genehunter-All, measures allele sharing in larger sets (Whittemore and Halpern, 1994; Kruglyak *et al.*, 1996). For a given inheritance vector (Kruglyak *et al.*, 1996), the Genehunter-All score function is defined as:

$$S_{all} = 2^{-a} \sum_{h} [\prod_{i=1}^{2f} b_i(h)!] \qquad \text{(Eq. 15.4)}$$

(a)

(b)

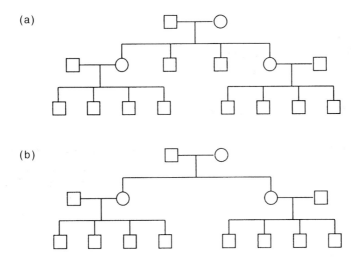

Figure 15.2 (a) Pedigree structure used for original simulated data (Goldin and Weeks, 1993). (b) Pedigree structure modified so that Genehunter pedigree size criterion is met

where a is the number of affected individuals in the pedigree, f is the number of founders, h is a collection of alleles obtained by choosing one allele from each of the affected individuals, $b_i(h)$ is the number of times that the ith founder allele appears in h (for $i = 1, \ldots, 2f$), and the exclamation point signifies the factorial. The Genehunter-All statistic should be more powerful in most cases than the Genehunter-Pairs statistic because it measures IBD sharing in sets of affected individuals while the 'pairs' score function only measures IBD sharing between pairs of relatives. For both Genehunter nonparametric linkage statistics a normal approximation is used to determine the statistical significance of an observed score. Furthermore, since the data do not always allow one to specify an inheritance vector with certainty, the score is taken as the expectation of the conditional inheritance distribution; Kruglyak *et al.* (1996) refer to the second approximation as the perfect-data approximation. In the presence of partially informative data, the perfect-data approximation results in a conservative statistic, which approaches the expected significance level as the information in the sample grows towards 100% (Kruglyak *et al.*, 1996).

Because Genehunter is limited to pedigrees that meet the $2n - f \le 16$ criterion (where n is the number of nonfounders and f is the number of founders; Kruglyak *et al.*, 1996), we had to remove two people from the original pedigree shown in *Figure 15.2a* ($2n - f = 2(12) - 4 = 20$). Therefore, for our power comparisons, we analyzed pedigrees with the structure given in *Figure 15.2b*

$(2n - f = 2(10) - 4 = 16)$. We compared power to detect linkage given by Genehunter-Pairs, Genehunter-All, SimIBD, and SimAPM. For the SimAPM and SimIBD statistics, we generated 1,000 simulated replicates to construct the empirical null distribution for each replicate in each disease model.

Effects of allele frequency misspecification

We simulated 1,000 replicates of 20 families typed at a single marker with two equally frequent alleles and unlinked to the disease locus. Each family had the structure given in *Figure 15.2a*. We then analyzed these data using SimIBD and SimAPM and report *p* values based on the conditional null distribution to examine the false-positive rates when allele frequencies were misspecified (see *Figure 15.4*). For comparison, we also analyzed the data with versions of SimIBD and SimAPM that computed empirical *p* values based on the unconditional null distribution.

False-positive rates

We examined the false-positive rates for the SimIBD and SimAPM statistics. We simulated 1,000 replicates of 20 families each (pedigree structure given in *Figure 15.2a*) with the marker locus unlinked to the disease locus in all the families. We then analyzed these data, calculating the false-positive rate as the observed proportion of replicates that were significant at $p \leq 0.05$ or $p \leq 0.01$. The expected proportion of significant statistics is equal to the *p* value at which the significance is assessed. A false-positive rate greater than the expected proportion of significant statistics would be considered non-conservative, while a false-positive rate less than the expected proportion would be considered conservative.

Results

Power to detect linkage

When comparing the power to detect linkage of Genehunter-All, Genehunter-Pairs, SimIBD, and SimAPM, we used the *p* values actually produced by the programs. Note that the *p* values for Genehunter-All and Genehunter-Pairs are conservative when the data are not fully informative (Kruglyak *et al.*, 1996). At a level of $p \leq 0.05$ (*Figure 15.3*), SimIBD is the most powerful test for linkage for

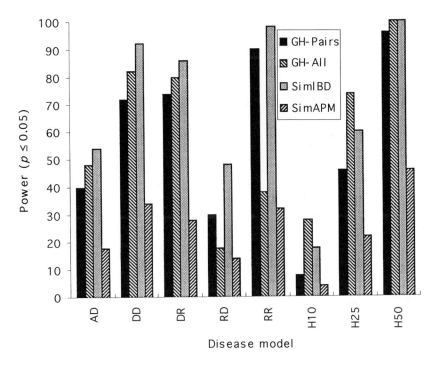

Figure 15.3 Power to detect linkage ($p \leq 0.05$) for SimAPM, SimIBD, Genehunter-Pairs, and Genehunter-All. See text for disease model codes

the AD, DD, DR, RD, and RR data sets. Genehunter-All is the most powerful statistic for H_{10} and H_{25} disease models. Both Genehunter-All and SimIBD had powers of 100% for the H_{50} disease model. Genehunter-All displays power that is between that of Genehunter-Pairs and SimIBD except for the RR and RD data sets where, interestingly, Genehunter-All is much less powerful than either Genehunter-Pairs or SimIBD. SimAPM exhibits much lower power than any of the other statistics for all disease models.

Effects of allele frequency misspecification

The data analyzed to produce *Figure 15.4* were originally simulated with the frequency of allele 1 set at 0.5, so the false-positive rate when analyses were done with the frequency of allele 1 set at 0.5 is very close to the expected level of 0.05. However, as the frequency of allele 1 varies from 0.5, the false-positive rates

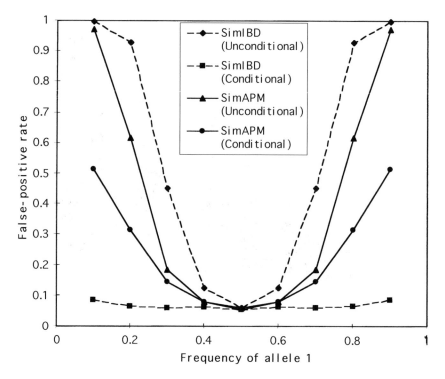

Figure 15.4 Effects of allele frequency misspecification on false-positive rate. Note that the original data were simulated with the frequency of allele 1 set to 0.5

increase, but note that the false-positive rates for the unconditional statistics are much higher than those for the conditional statistics. The observed differences between the conditional and unconditional simulation approaches are as expected: simulating conditional on the unaffected individuals' genotypes tends to 'correct' allele frequency misspecification toward the frequencies actually observed in the typed unaffected individuals. For the conditional statistics, SimAPM has a much higher false-positive rate than SimIBD. The relatively smaller effect of allele frequency misspecification on SimIBD than on SimAPM is expected because SimIBD, unlike SimAPM, uses the genotypes in the unaffected individuals (which do not change in our conditional simulation approach) to aid in inferring IBD status when computing its statistic. There is therefore a further buffering against elevation in the statistic over that provided by the conditional simulation itself.

False-positive rates

Measurement of false-positive rates (*Table 15.2*) revealed that the proportion of replicates that was significant for the SimIBD statistic (that is, the false-positive rate) was lower than expected. SimAPM had a false-positive rate that slightly exceeded the expected rate. Therefore, the SimIBD statistic was slightly conservative while SimAPM was slightly nonconservative.

Table 15.2. False-positive rates

Statistic	Expected false-positive rate	Observed false-positive rate
SimIBD	0.05	0.046
	0.01	0.009
SimAPM	0.05	0.060
	0.01	0.015

Timing results

Analyzing a single replicate (20 families of the structure given in *Figure 15.2b*) of the AD data set, Genehunter-Pairs took 32 min 40 sec (system + user time), Genehunter-All took 3 hr 4 min 8 sec, and SimIBD and SimAPM combined took a total of 4 min 15 sec (system + user time, 1,000 replicates in the simulated null distribution) running on our Sun SPARCstation 514MP. Other replicates and other disease models had comparable times. Note that analyzing more markers would not significantly increase the time requirements of either of the Genehunter nonparametric linkage statistics. However, because SimIBD and SimAPM must be run once for each additional marker, their time requirements would increase when analyzing more markers.

Discussion

Power to detect linkage

The data presented here show that our simulation-based statistic, SimIBD, is usually more powerful than the similar statistics, Genehunter-Pairs and

Genehunter-All, which use large-sample approximations to determine the p value. We had expected that the Genehunter-All statistic would provide the most powerful test for linkage in nearly every situation. However, SimIBD was more powerful for five of eight two-locus disease models presented here. Furthermore, the Genehunter-All statistic performs poorly with disease models with a strong recessive component (RR and RD disease models), but well with the heterogeneity models (H_{10}, H_{25}, and H_{50}) where both loci involved are dominant. A possible explanation for this may be that the Genehunter-All statistic is most powerful when all the affected individuals share the same allele IBD, which may happen often in a dominant disease, but less often in a recessive disease. For recessive diseases (and particularly for two-locus diseases), we would expect to see increased IBD sharing within each sibship, but not necessarily between the sibships. Genehunter-Pairs also has lower power than SimIBD for every disease model. Both Genehunter nonparametric linkage statistics employ the same approximation when determining the p value: it is probable that this approximation leads to the low power observed here; and this is consistent with the observation of Kruglyak *et al.* (1996) that their approximation-generated p values were conservative.

Multipoint analyses can improve power to detect linkage. We plan to implement a multipoint version of SimIBD to capitalize on this increased power. Genehunter already performs multipoint analyses, and including more markers should increase the power relative to that seen here for a single marker. Also, the Genehunter perfect-data approximation should progressively improve as more markers are included in the analysis and informativity increases, somewhat reducing the extremely conservative nature of the nonparametric linkage statistics from that observed here.

False-positive rates and effects of allele frequency misspecification

SimIBD and SimAPM are simulation-based statistics and therefore the false-positive rates are close to those expected at a given significance level (see *Table 15.2*). As we have chosen to use simulation conditional on the unaffected individuals' genotypes to construct the null distribution, the effects of allele frequency misspecification on the SimIBD and SimAPM statistics are reduced from those observed when unconditional simulation (gene-dropping) is used (see *Figure 15.4*). For SimIBD, the sensitivity to allele frequency misspecification is reduced further because genotypes in the unaffected individuals remain unchanged even after simulation, so even more information about the alleles in the original data is maintained when computing the null distribution.

Conclusions

The ASP study design represents one paradigm for mapping genes underlying complex diseases. However, with some diseases, it is more appropriate or desirable to allow families to be more complex than single nuclear families. Until recently, few methods existed for dealing with such complex extended families. The method of choice, the APM-method, has several disadvantages. Using a simulation-based methodology combined with a score statistic based on IBD rather than IBS sharing overcomes these difficulties and results in a very powerful statistic. Conditional simulation provides an excellent means of reducing the effects of allele frequency misspecification; SimAPM, and especially SimIBD, are robust to allele frequency misspecification. SimAPM and SimIBD are not limited to certain pedigree structures or sizes and can be applied equally well to sib pair data and large pedigrees, making them extremely versatile tools for detecting linkage.

The power of SimIBD compares very favorably with that of the Genehunter-All and Genehunter-Pairs statistics. In fact, SimIBD was the most powerful test for linkage in five of eight two-locus disease models tested here. The relatively low power of the Genehunter nonparametric linkage statistics may be due to the use of large-sample approximations that lead to an excessively conservative statistic (Kruglyak *et al.*, 1996) when the data are not fully informative. Intuitively, Genehunter-All could be expected to perform better than any statistic that deals with pairs of affected individuals only. In the current implementation that uses both the normal approximation and the perfect-data approximation, both Genehunter nonparametric linkage statistics appear very conservative and could perhaps benefit from an option to compute simulation-based empirical p values.

Using SimIBD and SimAPM

SimIBD and SimAPM are both computed using the simibd package, which can be obtained by anonymous file transfer protocol from `watson.hgen.pitt.edu` or `ftp.ebi.ac.uk`. The package, simibd, requires two LINKAGE-format input files, a pedigree file (after processing with makeped), and a locus file. The trait locus must be of the type affection status; marker loci must be codominant and may be coded in numeric or binary factor form.

Acknowledgments

This work was supported by NIH grant HG00719, the University of Pittsburgh, the Wellcome Trust Centre for Human Genetics at the University of Oxford, the Association Française Contre Les Myopathies (AFM), the WM Keck Center for Advanced Training in Computational Biology at the University of Pittsburgh, Carnegie Mellon University, and the Pittsburgh Supercomputing Center.

References

Cottingham, R.W., Jr., Idury, R.M., Schaffer, A.A. Faster sequential genetic linkage computations. (1993) *Am. J. Hum. Genet.* **53 (1):** 252–263.

Davis, S., Schroeder, M., Goldin, L.R., Weeks, D.E. Nonparametric simulation based statistics for detecting linkage in general pedigrees. (1996) *Am. J. Hum. Genet.* **58 (4):** 867–880.

Goldin, L.R., Weeks, D.E. Two-locus models of disease: comparison of likelihood and nonparametric linkage methods. (1993) *Am. J. Hum. Genet.* **53 (4):** 908–915.

Kruglyak, L., Daly, M.J., Reeve Daly, M.P., Lander, E.S. Parametric and nonparametric linkage analysis: a unified multipoint approach. (1996) *Am. J. Hum. Genet.* **58 (6):** 1347–1363.

Lander, E.S., Schork, N.J. Genetic dissection of complex traits. (1994) *Science* **265 (5181):** 2037–2048.

Martinez, M., Goldin, L.R. Power of the linkage test for a heterogeneous disorder due to two independent inherited causes: a simulation study. (1990) *Genet. Epidemiol.* **7 (3):** 219–230.

Ott, J. Computer-simulation methods in human linkage analysis. (1989) *Proc. Nat. Acad. Sci. USA* **86 (11):** 4175–4178.

Risch, N. Linkage strategies for genetically complex traits. III. The effect of marker polymorphism on analysis of affected relative pairs. (1990) *Am. J. Hum. Genet.* **46 (2):** 242–253.

Weeks, D.E., Lange, K. The affected-pedigree-member method of linkage analysis. (1988) *Am. J. Hum. Genet.* **42 (2):** 315–326.

Weeks, D.E., Lange, K. A multilocus extension of the affected-pedigree-member method of linkage analysis. (1992) *Am. J. Hum. Genet.* **50 (4):** 859–868.

Weeks, D.E., Ott, J., Lathrop, G.M. (1990) SLINK: a general simulation program for linkage analysis. (Abstract) *Am. J. Hum. Genet.* **47:** A204.

Whittemore, A.S., Halpern, J. A class of tests for linkage using affected pedigree members. (1994) *Biometrics* **50 (1):** 118–127.

SECTION IV

<u>TOOLS</u>

Chapter 16
The Human Genome Mapping Project Resource Centre Computing Service
Frank R. Visser

Introduction

The Human Genome Mapping Project Resource Centre (HGMP-RC) was founded in 1990 by the UK Medical Research Council. It is based at the new Wellcome Trust Genome Campus in Hinxton near Cambridge, UK. The campus houses the HGMP-RC, the Sanger Centre and the European Biomolecular Institute (EBI).

The HGMP-RC was started primarily for the benefit of users in the UK, but in the past few years it has attracted users from Europe, the USA and the rest of the world. Use of the HGMP-RC services is free for academic users, but commercial users have to pay a fee.

The HGMP-RC was established to provide resources to the genetic community involved in projects mapping the genomes of different organisms, but with the primary focus on the Human Genome Project. The HGMP-RC offers two different services, biological services and computational services. A short summary of the biological services will be given, but this chapter will concentrate mainly on the computational, and more especially on the linkage and mapping services.

Biological Services

The HGMP-RC offers the following biological services:

GENETIC MAPPING OF DISEASE GENES
ISBN 0-12-232735-7

- cDNA and Genomic Libraries: these are used for cloning specific genes.
- Yeast artificial chromosome (YAC) libraries: these are used for cloning large DNA segments. For example, recently a YAC contig map covering 75% of the human genome was publicised.
- P1 artificial chromosome (PAC) libraries: these are derived from the bacteriophage P1 and the F factor in *Escherichia coli*, and are used like YACs for cloning smaller (100–150 kb) DNA fragments.
- Cosmid libraries: these have the smallest insert size for genomic cloning (about 40 kb). Cosmid libraries containing the genome of the fugu (*Fugu rubripes*) or puffer fish are available.
- CpG island libraries: CpG islands are short stretches of DNA containing a high density of non-methylated CpG dinucleotides. These islands are associated with coding regions, so a CpG island library is a genomic library that has been enriched with coding regions.
- Hybrid panels: these are used for mapping and localisation of markers.
- UK DNA Probe Bank: this contains about 1,000 probes for screening by hybridisation.
- Primer Bank and oligonucleotide synthesis: this is a continuously expanding collection of primers used for genetic analysis by polymerase chain reaction (PCR).
- Comparative mapping: the use of 1,000 mouse backcrosses allows mapping of the mouse genome with an average distance of 0.3 centimorgan (cM) between the markers. The aim is to have about 4,000 markers across the mouse genome.

More information on each of these services can be found on the World Wide Web (WWW). The Uniform Resource Locater (URL) address is: `http://www.hgmp.mrc.ac.uk/Public/Docs/Bio/Bio.html`

Research

The biological service participates in several research projects. It cooperates with the French Pasteur Institute in the European Collaborative Interspecific Backcross (EUCIB) Project (Breen *et al.*, 1994). The primary goal of EUCIB is the creation of a high-resolution map of the mouse genome. This map will be useful for comparative mapping. This project is now nearly finished. More information can be found on the WWW at `http://www.hgmp.mrc.ac.uk/MBx/MBxHomepage.html`

A second large project is the fugu (*Fugu rubripes*) project: the creation of a landmark map of the fugu (puffer fish). The fugu fish has essentially the same number of genes as the human genome, although its genome size is only approximately 400 Mb against 3,000 Mb in most mammals. Data obtained to date are consistent with an overall eight-fold compression of the fugu genome compared to that of the human. Isolation of fugu homologies to human genomic regions can therefore facilitate gene finding in the human genome (Brenner *et al.*, 1993). More information can be found on the WWW at `http://fugu.hgmp.mrc.ac.uk/fugu/fugu.html`

Computational services

The computational service was started in the knowledge that the amount of information and programs available on the Internet had become so large that it was virtually impossible for the individual researcher to keep track of all or even most of it. The HGMP-RC sees its role as firstly to collect all the relevant information and programs in one place, and secondly to shield the user as much as possible from the computer by using a menu-based interface. The interface has recently been ported to the WWW at `http://www.hgmp.mrc.ac.uk/` This makes access easier for the majority of our users, and presents the user with a familiar interface.

The computational service of the HGMP-RC has enjoyed a healthy growth over the last five years. The current (July 1996) number of registered (computational and non-computational) users stands at about 5,000. The growth has been rapid, with a doubling of the number of users in less than two years. *Table 16.1* shows the geographical distribution of our computer users. Most users are in the UK, but around 15% are from the EU countries.

The HGMP-RC provides training courses consisting of general courses for users to familiarise themselves with the systems, and more specialised courses focused on a program or a group of related programs. The courses are:

- HGMP General Computing Course.
- Disease Mapping.
- Map Construction.

Table 16.1. Location of registered users

Location	Number
UK	3557
EU	461
Other	222
Commercial	17
Total	4257

- Getting Information from the Network.
- An Introduction to the Genome Database (GDB).
- A *Caenorhabditis elegans* Database (ACeDB) Workshop.
- Sequencing Project Management using the Staden Package.
- GCG (Genetics Computer Group) Software Gene Identification and Prediction.
- Protein Structure Prediction.
- Phylogenetic Trees from Molecular Sequences.

More information and registration forms for the courses can be found on the HGMP-RC WWW site at
`http://www.hgmp.mrc.ac.uk/Public/ Docs/Courses/courses.html`

Access, Interfaces and User Support Helpdesk

An essential part of the service is the User Support Helpdesk. It can be contacted via e-mail (`support@hgmp.mrc.ac.uk`), telephone (+44 (0)1223 494520) or fax (+44 (0)1223 494516). All user queries are registered and dealt with by the appropriate person for that particular query. Periodically the User Support Group is informed about which queries are still outstanding. More than half of the queries are answered the same working day.

The HGMP-RC believes it is essential to keep in touch with the user community to determine their main problems and needs. There is therefore a need to communicate with the users. The main channels of communication are user support queries. All users are therefore encouraged to contact the helpdesk not only for problems, but also for information and suggestions for new programs, databases or services. Members of the user support staff regularly attend meetings and symposia to keep in touch with the wider biological community in the UK and Europe.

A major challenge is providing an easy to use interface to all the programs available on our systems. The HGMP-RC uses several different Unix operating systems, for example AIX (IBM) for the LINKAGE programs and Solaris (Sun) for most other programs. It is also necessary to divide the load of the programs as evenly as possible over the different computers at the HGMP-RC. The user support staff have developed a menu system to shield all this complexity from the user. The menu can be accessed via a simple telnet connection to the HGMP-RC. A program or database can be selected via the menu, and the menu system starts the program on the correct machine and presents the user with a command window. A large effort has been undertaken by the user support staff to port the menu system to the WWW, providing users with a 'point and click' interface. This version of the menu came on-line in September 1995, and works extremely well. Many databases and programs have started to provide a WWW forms-based interface. This makes the new menu system and the program or database interfaces significantly more consistent and therefore easier to use. It also avoids users being confronted with the underlying Unix operating system or difficult to learn command-line interfaces.

There are still many programs that do not have a WWW forms interface, especially the LINKAGE programs. This makes the programs quite difficult to use, and a knowledge of Unix is needed to use the programs effectively. The HGMP-RC is actively involved in creating easy to use user interfaces for these programs.

Available software and databases

The computational services can be roughly divided into the following groups, excluding linkage analysis.

- Genome databases.
- Sequence analysis and databases.
- Mapping analysis.
- Bibliographic services.
- Miscellaneous.

However, the differences between the groups are not well defined: genome databases contain DNA sequences and could therefore be classified as sequence databases. In the same way, most sequence databases provide (part of) the journal articles in which a sequence is described, and could therefore be classified under bibliographic services.

Genome Databases The HGMP-RC manages and provides links to many databases with genomic data from dozens of species. The most important are the human genome databases, and the most important of these is Genome Database (GDB) (Fasman *et al.*, 1996), which holds data on human gene loci, polymorphisms, mutations, probes, genetic maps, GenBank, citations and contacts. With the newly released version (Version 6.0), GDB has become the first database of its kind to allow on-line public curation and third party annotation. Another important database is the online version of McKusicks Mendelian Inheritance In Man (OMIM) (Rashbass, 1995), which is a catalogue of human genes and known genetic disorders. The HGMP-RC provides links to chromosome-specific databases, mutation databases and many other more specialised databases.

211

The HGMP-RC hosts many non-human databases. Several are specialised databases for the mouse and rat genomes, which are important model organisms. Examples are the Gene Knockout Database and the Gene Expression Database.

Another important database is AceDB (A *Caenorhabditis elegans* Database). The front end of this database is based on X-Windows, and is used by a number of other genomic databases including the Integrated Genome Database (IGD), (Ritter *et al.*, 1994), AtDB (A *Arabidopsis thaliana* Database) and MycDB (A *Mycobacterium* Database).

Databases that integrate genomic and metabolic information are an interesting new development. A first example is EcoCyc (Karp *et al.*, 1996b), which contains genetic and metabolic data on *E. coli*. A new database, HinCyc (*Haemophilus influenzae*) in the same format has been released recently (Karp *et al.*, 1996a). It is clear that in the future genomic databases will integrate genomic and metabolic data.

Sequence Analysis and Databases Sequence analysis is the basis of all genetic research. Information gathered from sequence analysis of genes can lead to indications of the structure of the protein, its probable function and sometimes even information on its probable three-dimensional conformation. The sequences in or near the promotor areas determine in which circumstances the gene will be expressed. The databases containing sequence and protein data have become extremely large, and are set to grow significantly in the coming years. However, the value of a database for research is not so much determined by the amount of sequence data in it, but by the annotation and error checking performed on the data. This is the most labour intensive task and difficult to automate. Intensive research is undertaken to develop, for example, faster searching algorithms and better interfaces for users.

The most comprehensive databases can be divided into two groups: firstly pure sequence databases, such as GenBank (Benson *et al.*, 1996) and the EMBL database, which contain all sequence data available, and secondly protein databases, such as SWISS-PROT (Bairoch and Apweiler, 1996) and Protein Information Resource (PIR) (George *et al.*, 1996), which concentrate mainly on protein sequences. SWISS-PROT is in the opinion of most experts the most valuable database because of the excellent annotation and rigorous checking of possible errors in the sequences. As a consequence, SwissProt is the smallest sequence database. To keep the annotation to the current level with the enormous mass of new data will be difficult in the coming era of massive sequencing projects.

As with the genome databases, a host of smaller and more specialised databases exist. Examples include the following.

- dbEST, a database that contains sequence and mapping data on 'single-pass' cDNA sequences or expressed sequence tags from a variety of organisms.

- A Transcription Factor Database (TFD).
- The Tumor Gene Database.
- The Eukaryotic Promotor Database (EPD).

These examples are merely a small sample of databases available through the HGMP-RC. New databases are coming on-line regularly, and the HGMP-RC is adding the new databases to its menu system as soon as they come available.

Tools for accessing and searching sequence databases are under active development. An important recent program in this regard is Sequence Retrieval System (SRS) (Etzold and Argos, 1993). SRS provides a common WWW-based interface to many different sequence and genomic databases. It provides the user with easy to use search and comparison tools. Because it provides a united and consistent interface, searching many different databases becomes an almost trivial task.

Bibliographic Services Because of the explosive growth in available data, it has become essential for biology researchers to have on-line access to the latest information. The HGMP has therefore a wide range of bibliographic services available. The facility having the largest collection of bibliographic information is Bibliography Information Database System (BIDS). BIDS is not one database, but gives access to the following three separate databases.

- ISI (Institute for Scientific Information) Science Citation Index Database, which is the machine-readable version of the printed Science Citation Index with additional journal coverage from the Current Contents series of publications. It contains bibliographic references (but not abstracts) of articles, reports, papers, discussions, editorials, notes, reviews and other items from over 4,400 journals in the field of natural, physical and biomedical science and technology. Each year contains over 700,000 articles.
- Excerpta Medica Database (EMBASE), which covers about 3,500 biomedical journals from 110 countries and specialises in the areas of drugs and toxicology. About two-thirds of the articles include abstracts. Where the original article has a non-English abstract, the abstract is not included, but the title is included in both English and the original language.
- British Library Inside Information (BLII), which covers details of every major article in some 10,000 journals, with a large international coverage over a wide range of subject areas. A particular strength of the data is that they are current, frequently updated, and quickly available on-line. The data are supplied by the British Library Document Supply Centre. New journal issues are received at the Centre soon after publication, and the computer-produced records are ready within a few working days. The BLII data are

updated twice a week, allowing fast access to newly published material. Approximately one million articles (including letters and relevant editorials) are added to the file each year. The collection of data was commenced in October 1992.

Entrez is an important bibliographic research tool. It provides an integrated approach for gaining access to nucleotide and protein sequence databases (GenBank, PIR and SWISS-PROT) and to the MEDLINE citations in which the sequences were published. With Entrez, the user can rapidly search several hundred megabytes of sequence and literature data using techniques that are quick and intuitive.

A key feature of Entrez is the concept of 'neighbouring' information, which relates records within the same database by statistical measurements of similarity. This concept makes it easy for a user to locate related references or sequences by asking Entrez to 'Find all papers that are like this one' or 'Find all sequences that are like this sequence.' This is extremely useful for finding new relationships between already-known data, and opens up new and unexpected connections.

In addition to neighbouring information, Entrez also has the concept of 'hard links,' which are specific connections between entries in different databases. For example, for each MEDLINE article in the Entrez database, there are hard links to any protein or nucleotide sequences that were published in that article, and the cited protein or nucleotide sequences have reciprocal hard links back to the MEDLINE articles. Nucleotide sequences and the proteins derived from them by conceptual translation also have hard links to one another. This means that starting with an article that publishes a sequence, it is possible to go to the actual sequence in the database and find neighbouring sequences, which can lead to new insights on the probable function or evolutionary relationships of this sequence.

The HGMP-RC provides links to on-line journals. Some journals provide only a table of contents and abstracts (e.g. *Nature*), some provide full access to the full articles (e.g. *British Medical Journal*), and others provide full access to registered subscribers (e.g. *Nucleic Acid Research*).

Linkage analysis at the HGMP-RC

Linkage analysis is an active area of research, especially on the computational side. Many articles are published announcing linkage of genes with common or not so common diseases. However, for more complex diseases such as schizophrenia, linkage analysis has proved to be a frustrating experience in the past decade. For more complex, multigenetic and strongly environmentally influenced diseases and behaviours the classical method of linkage analysis becomes difficult. New more powerful methods are under intense scientific scrutiny.

The HGMP-RC is committed to actively supporting and maintaining linkage analysis programs and the linkage analysis community in the UK and Europe. One member of the user support team is solely dedicated to supporting the linkage analysis packages found on the HGMP-RC's computer systems. The HGMP-RC's activities can be divided into the following groups.

- Programs: the HGMP-RC has the largest collection of software for linkage analysis available anywhere. This collection is kept up-to-date and is actively supported.
- Documentation: a large effort has been undertaken to get most of the program documentation available on the WWW. The HGMP-RC is also participating in projects to provide biologists with a useful introductionary document to linkage.
- Courses: the HGMP-RC has been providing courses for linkage analysis, which are currently under review and will be coordinated with the new Linkage Hotel project.
- User interfaces and data management: the HGMP-RC is closely involved with research groups developing new user interfaces to interact with the linkage programs.

Linkage Programs Available at the HGMP-RC

The HGMP-RC supports about 30 linkage analysis packages, which are divided into four submenus.

General linkage analysis menu option These programs are the most often used and the most popular is FASTLINK. The original version of the program was called LINKAGE (Lathrop *et al.*, 1984) and was written for personal computers using pascal. It was adapted for large Unix-based computers and converted into C, and in the process renamed to FASTLINK (Cottingham *et al.*, 1993). Several speed-ups have been added by Professor Schaffer (Schaffer, 1995). The package consists of the programs ILINK and MLINK (for two-point linkage analysis), and LINKMAP (for multipoint analysis). An excellent book is available describing how to use the FASTLINK programs (Terwilliger and Ott, 1994). The version currently installed is 3.0. The HGMP-RC is looking at installing the parallel version of FASTLINK. A problem with large scale multipoint analysis is that it requires an enormous amount of computer power before results are generated. Several schemes for speeding up have been proposed, but the most successful seems to be VITESSE (O'Connell and Weeks, 1995), which is compatible with the FASTLINK package. The new algorithm speeds up the calculation significantly. Another, less often used multipoint analysis program is GAS, which provides an alternative interface to the VITESSE package.

Other important packages on this menu are several linkage simulation programs such as SIMLINK (Boehnke, 1986) and SLINK. Before committing to a large LINKMAP run, these programs can be used to test if there is enough power in the pedigree information for a full scale multipoint analysis.

Many programs have been released in the previous few years based on nonparametric methods. Examples are the affected pedigree member (APM)-method, SPLINK, and SimIBD.

Linkage Map Construction Programs The difference between linkage programs and mapping programs is somewhat arbitrary, but necessary to minimise the number of choices on each menu. Examples are CRI-MAP (Lander and Green, 1987) and the programs developed by the Whitehead Institute (MAP-MAKER, SIBS (Kruglyak *et al.*, 1995) and HOMOZ (Kruglyak and Lander, 1995)). Radiation hybrid mapping programs such as RHMAP are also located here.

Miscellaneous Linkage Programs The programs here do not fit into the previous two categories, but are still related to linkage analysis. Important examples are SIMWALK and SIMCROSS (Weeks and O'Connell, 1995), two haplotyping programs, HOMOG and MTEST for calculating heterogeneity, and NOCOM for mixture analysis.

Linkage Utility Programs Some of the programs in this menu provide user interfaces to the linkage programs and data management tools. Examples of these are QDB/DOLINK, and IGD/XPED (Spiridou *et al.*, 1996). IGD/XPED is relatively new and based on the ACeDB database backend. Both these programs require X-Windows and are therefore only useful for a relatively small group of people. Efforts are under way to provide interfaces based on the WWW. Other programs are PedPack, a pedigree drawing package, and ANALYZE, which simplifies testing for linkage.

Documentation

The HGMP-RC has made an effort to provide useful documentation with all programs installed on its systems. For most programs, documentation and example files are provided and are available on the WWW. Efforts are underway to provide documents to introduce users to linkage, give advice on which programs to use, and point out problems and pitfalls researchers can encounter when starting with linkage analysis.

Courses

The HGMP-RC has provided introductory courses for linkage analysis since 1990 aimed at giving the users a grounding in linkage analysis and making them more familiar with the programs and tools available.

Currently the courses are under review to coordinate them with new initiatives and make them more up-to-date with current needs. The plans as they are now under review are to provide three different kinds of course: a simplified introductory course, a more comprehensive overview course, and specialist seminars where new methodologies will be reviewed and practical problems discussed.

The aim of the new approach is to create a group of professional linkage experts who will cooperate with biologists to assess their data and review projects. One of the new initiatives is the Linkage Hotel. It will start operating in early 1997 and has two main aims.

- To provide a centralised resource for linkage mapping in a 'Hotel' context.
- To provide integrated training courses covering the biochemical, instrumentational and computational aspects of the methodology.

The facilities will emulate the Genethon model, which provides dedicated centralised mapping facilities for visiting workers to be trained in microsatellite-based linkage mapping under the guidance of expert staff. Most laboratories have neither the capital equipment nor expertise to carry out these studies themselves, but will be enabled by the hotel facility to cross the boundary between phenotype and physical map and to proceed with molecular investigations. The projects will be accompanied by courses to give the users the necessary background information on the computational aspects of linkage analysis.

Future developments in the linkage facilities

The computing facilities are evolving rapidly, as is the whole field of linkage analysis. The HGMP-RC is concentrating its efforts on making the linkage programs easier to use by providing better interfaces, mainly based on the WWW, to the programs available. The HGMP-RC is looking into the possibility of using the new WWW based language Java$^{(TM)}$ to create a common interface to the linkage programs based on a graphical, pedigree-centred view of linkage analysis. The HGMP-RC aims to concentrate data entry and data maintenance on its systems and not on local systems without users having to invest in new software such as *X* servers. Users running Windows with a Java$^{(TM)}$-enabled WWW browser will be able to use it.

The HGMP-RC hopes to integrate this program (when it is functional) with currently available data management tools such as IGD/XPED to ease the setting up of new linkage analysis, and centralise the access to programs through this interface. This means that users have more time to spend on grasping the functionality of the programs instead of struggling with details such as formating input files and understanding the format of the output files. It is hoped that authors of new programs will recognise that integrating their programs with this interface will speed up development and free them to concentrate on the algorithms instead of the user interface.

Creating and maintaining useful linkage documentation will be one of the main efforts of the HGMP-RC's user support. It will be coordinated with experience gathered from the 'Linkage Hotel' facility so that the larger linkage community will be able to profit from this new initiative.

References

Bairoch, A., Apweiler, R. The SWISS-PROT protein sequence data bank and its new supplement TREMBL. (1996) *NAR.* **24 (1):** 21–25.

Benson, D.A., Boguski, M., Lipman, D.J., Ostell, J. GenBank. (1996) *NAR.* **24 (1):** 1–5.

Boehnke, M. Estimating the power of a proposed linkage study: a practical computer simulation approach. (1986) *Am. J. Hum. Genet.* **39 (4):** 513–527.

Breen, M. Towards high resolution maps of the mouse and human genomes – a facility for ordering markers to 0.1 cM resolution. European Backcross Collaborative Group. (1994) *Hum. Mol. Genet.* **3 (4):** 621–627.

Brenner, S., Elgar, G., Sandford, R., Macrae, A., Venkatesh, B., Aparicio, S. Characterization of the pufferfish (*Fugu*) genome as a compact model vertebrate genome. (1993) *Nature* **366 (6452):** 265–268.

Cottingham, R.W. Jr, Idury, R.M., Schaffer, A.A. Faster sequential genetic linkage computations. (1993) *Am. J. Hum. Genet.* **53 (1):** 252–263.

Etzold, T., Argos, P. SRS – an indexing and retrieval tool for flat file data libraries. (1993) *Comput. Appl. Biosci.* **9 (1):** 49–57.

Fasman, K.H., Letovsky, S.I., Cottingham, R.W., Kingsbury, D.T. Improvements to the GDB human genome data base. (1996) *NAR.* **24 (1):** 57–63.

George, D.G., Barker, W.C., Mewes, H.W., Pfeiffer, F., Tsugita, A. The PIR-international protein sequence database. (1996) *NAR.* **24 (1):** 17–20.

Karp, P.D., Ouzounis, C., Paley, S. (1996a) HinCyc, a knowledge base of the complete genome and metabolic pathways of *H. influenzae* in *Intelligent Systems for Molecular Biology 1996*. AAAI Press, St. Louis, MO.

Karp, P.D., Riley, M., Paley, S.M., Pelligrini Toole, A. EcoCyc: an encyclopedia of *Escherichia coli* genes and metabolism. (1996b) *NAR*. **24 (1):** 32–39.

Kruglyak, L., Daly, M.J., Lander, E.S. Rapid multipoint linkage analysis of recessive traits in nuclear families, including homozygosity mapping. (1995) *Am. J. Hum. Genet.* **56 (2):** 519–527.

Kruglyak, L., Lander, E.S. Complete multipoint sib-pair analysis of qualitative and quantitative traits. (1995) *Am. J. Hum. Genet.* **57 (2):** 439–454.

Lander, E.S., Green, P. Construction of multilocus genetic linkage maps in humans. (1987) *Proc. Nat. Acad. Sci. USA.* **84 (8):** 2363–2367.

Lathrop, G.M., Lalouel, J.M., Julier, C., Ott, J. Strategies for multilocus linkage analysis in humans. (1984) *Proc. Nat. Acad. Sci. USA.* **81 (11):** 3443–3446.

O'Connell, J.R., Weeks, D.E. The VITESSE algorithm for rapid exact multilocus linkage analysis via genotype set-recoding and fuzzy inheritance. (See comments) (1995) *Nat. Genet.* **11 (4):** 402–408.

Rashbass, J. Online mendelian inheritance in man. (1995) *Trends. Genet.* **11 (7):** 291–292.

Ritter, O., Kocab, P., Senger, M., Wolf, D., Suhai, S. Prototype implementation of the integrated genomic database. (1994) *Comput. Biomed. Res.* **27 (2):** 97–115.

Schaffer, A.A. Faster linkage analysis computations for pedigrees with loops or unused alleles. (1996) *Hum. Hered.* **46 (4):** 226–235.

Weeks, D.E., Sobel, E., O'Connell, J.R., Lange, K. Computer programs for multilocus haplotyping of general pedigrees. (1995) *Am. J. Hum. Genet.* **56 (6):** 1506–1507.

Chapter 17
The display of comparative mapping data using ACEDB

Jo Dicks

Introduction

Chromosomal evolution in mammals is conservative, in that the chromosomal mutations that have occurred during the evolution of these species are rare events. Consequently the chromosomes of the mammals are very similar in terms of the genes located along them. If a chromosome is considered to be a linear array of genetic markers, the main difference between chromosomes of different species comes in the ordering of those markers, this having changed as the species have evolved from their closest common ancestor. Although the ordering has changed, blocks of genes that are close in one species can be seen to be close in other closely related species. These blocks are known as 'conserved segments' of genes. Conserved segments of genes are not unique to the mammals and a similar situation may be seen in other groups of related species.

When two genes, one in each of two species, are related by evolution from a common ancestor such that they carry out a similar function in each of the two species, or that they have a similar amino acid content, then they are known as *homologous* to one another. Comparative mapping studies look at homologous data and attempt to map genes in one species by looking at the positions of their homologues in another. This type of data is becoming more abundant, so to get a basic idea of the differences or similarities of gene ordering in two species simple diagrams are needed. Such diagrams provide information on the nature of

GENETIC MAPPING OF DISEASE GENES
ISBN 0-12-232735-7

evolutionary change and also allow data from one species to validate inferences on order in another.

The usefulness of comparative mapping in the field of mapping and its ability to aid the understanding of evolutionary changes of the chromosomes has become recognised as more genetic maps are made. There are currently over 100 species with genetic maps, and over 30 of these are mammals. Other genetic maps are of species belonging to groups as diverse as grasses, flies, bacteria, and viruses. For a review of the recent status of comparative mapping projects see Edwards (1994) and O'Brien (1991).

An example of comparative mapping is to imagine three species (perhaps three mammals or three grasses) connected in time by a fairly recent ancestor. A short conserved segment of an ancestral chromosome is preserved in each of these species, on chromosome 1 in species A, chromosome 3 in species B, and chromosome 10 in species C (*Figure 17.1*). A particular gene is known to be located within the conserved segment in species A and its homologue in species B is within the conserved segment in that species. The gene is unmapped in species C. However, we can hypothesise that the gene is located on chromosome 10 in species C within the conserved segment. Although we are unable to say our hypothesis is correct, we can make the hypothesis with a fair degree of confidence, and we have a starting point for mapping the gene in species C.

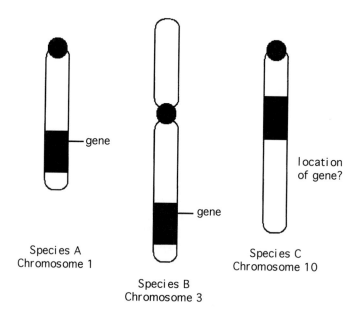

Figure 17.1 An example of the use of comparative mapping

A recent project funded by the Medical Research Council (MRC) aimed to produce a comparative mapping database that could aid such interspecies analyses using the database software ACEDB, it's name being derived from the species database ACeDB (*Caenorhabditis elegans* Database). This is a database rich in graphical functionality, originally developed for the nematode genome project by Richard Durbin and Jean Thierry-Mieg (1991). It was written in C for use on UNIX workstations. Apple Macintosh and Windows versions were later adapted from it. It is extremely simple to use so the user can begin almost immediately, but it takes time to learn how to use it comprehensively. The data held can be of many types, for example physical mapping data, sequence data and references. The system is navigated by using the mouse, through a series of windows that contain either a graphical interpretation of the data or text information. For example, an object such as a gene or a chromosome may be selected by the mouse and all information relevant to that object can be brought up on screen. The system has its own query language and the displayed data can be restricted by a particular query if required. The database can also be edited during use.

Production of the comparative mapping version of ACEDB was carried out in collaboration with the Mammalian Genetics Unit in Harwell, Oxon. A key consideration of the project was to develop new graphics tools to show how two or more species differ genetically. New methods of comparing species within this version of ACEDB include the following.

- The Oxford Grid.
- The Pairwise Chromosome Map.
- The One-to-Many Chromosome Map.
- The Species Grid.
  These maps are described in detail below.

The Oxford Grid

One useful method of comparing any two species is the Oxford Grid (Edwards, 1991). This term was originally applied by V. A. McKusick to a diagram in the second of the series of papers from Harwell and Oxford on human–mouse homologies (Dalton *et al.*, 1981; Buckle *et al.*, 1984; Searle *et al.*, 1987, 1989), and has been represented by figures in recent editions of *Mendelian Inheritance in Man* (McKusick, 1988).

The grid consists of a square. One species is allocated to the vertical axis and the other to the horizontal axis. Each side of the square is divided up into parts representing the chromosomes of these species. Gridlines are drawn across the square, joining up the ends of the chromosomes. Therefore the columns of the grid represent the chromosomes of the first species and the rows represent the chromosomes of the second species. The chromosomes in every species vary in size and the widths of the columns and rows vary accordingly. Points are plotted within the grid, each point representing a homologue between the two species. *Figure 17.2* shows a typical screen shot of the comparative maps in ACEDB. If you look at the Oxford Grid within this Figure you will see five spots in the highlighted 3,14 cell. This means that there are five genes on human chromosome 3 that are known to have homologues on mouse chromosome 14. Note that points are plotted at random within the cells and not according to their location along the chromosome. The resolution of this map is not great enough to gain any useful information by doing this.

An essential feature of the Oxford Grid is that the areas of the cells would have an equal density of occupancy by loci if there were no conserved segments. Therefore cells that are densely populated with points may indicate the presence of one or more conserved segments between the two relevant chromosomes. The grid can also be used to show weak homologies. The abundance of known homologous loci is now so high (over 1,000 in human and mouse) that any cell occupied by only one locus must be regarded as suspect. The identification of spurious homologies is an important use of the Oxford Grid. For example, the ordering of loci in the human by genetic methods cannot achieve the reliability or precision in the mouse and a locus may be assigned incorrectly to a particular chromosome.

The Oxford Grid is the most important new map in the comparative mapping version of ACEDB. It may be used to look for particular subsets of data. For example, Table 17.1 shows the human–mouse homologies known to be involved in eye disease. *Figure 17.3* shows an ACEDB Oxford Grid of these loci.

The Oxford Grid may aid comparative evolutionary studies of related species. For example, *Figure 17.4* shows three Oxford Grids comparing loci in human with their known homologues in mouse, cow, and chimpanzee. The differences between the three grids are striking, even allowing for the different numbers of loci in each grid. Although it has been suggested that chromosomal evolution, unlike gene evolution, does not proceed in a stochastic manner (i.e. it does not follow a probability model, particularly one which exhibits a time element) it is a general rule that those species with similar karyotypes are more closely related than those with widely differing karyotypes. It is simple to infer from the grids that of the three species, human is most closely related to the chimpanzee, as is known to be the case. It would not be trivial, however, to decide which of mouse and cow is the closer relative, but a general picture of the chromosomal evolution of these species is given.

The Oxford Grid may also be used as a paralogy map. When a gene in a

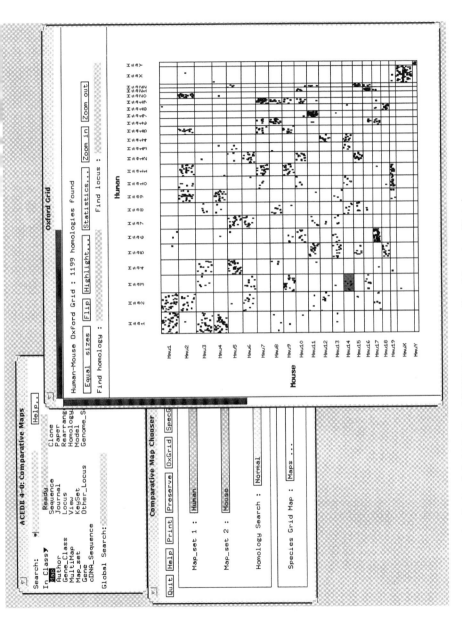

Fig. 17.2 An Oxford Grid in the ACEDB Comparative Mapping Database

Table 17.1. Human–mouse homologies involved in eye disease

Locus name	MIM	Human	Human bands	Mouse	Mouse location	Mouse bands
Albinism, oculocutaneous type II	203200	P	15q11.2-q12	p*	7.27	B5-D3
Albinism, tyrosinase negative	203100	TYR	11q14-q21	Tyr*	7.41	D3
Aniridia-2	106210	PAX6	11p13	Pax6*	2.49	E3-E4
Cataract, Coppock-like	123660	CRYGA	2q33-q35	Cryg1	1.31	C2-C4
Ectopia lentis	129600	FBN1	15q21	Fbn1	2	F
Fabry	301500	GLA	Xq22	Ags	X.58	F1
Galactokinase	230200	GALK1	17q21-q22	Glk	11.77	E1-E2
Galactosaemia	230400	GALT	9p13	Galt	4.19	A1-A5
Gangliosidosis, generalised GM1	230500	GLB1	3p23-p22	Bgl	9.62	E3-ter
Homocystinuria, B6 responsive	236200	CBS	21q22.3	Cbs	17.18	A3-B
Incontinentia pigmenti	308300	IP1	Xp11.21-cen	Td*	X.03	A1-A2
Krabbe	245200	GALC	14q21-q31	twi*	12.29	cen-F1
Norrie	310600	NDP	Xp11.23	Ndp	X.03	A1-A2
Nance–Horam	302350	NHS	Xp22.3-p21.1	Xcat	X	F3-ter
Ornithinaemia, gyrate atrophy	258870	OAT	10q26	Oat	7.64	F1-F4
Peters' anomaly	261540	PAX6	11p13	Pax6*	2.49	E4-F3
Phenylketonuria, common form	261600	PAH	12q22-q24.2	Pah*	10.53	C2-D1
Piebaldism	172800	KIT	4q12	Kit*	5.37	D-E5
Retinoblastoma	180200	RB1	13q14.2	Rb1*	14.27	D3-E2
Retinitis pigmentosa, rhodopsin	180380	RHO	3q21-q24	Rho	6.49	F
Retinitis pigmentosa, peripherin	179605	RDS	6p12	Rd2*	17.24	D-E1
Retinol binding protein-1	180260	RBP1	3q21-q22	Rbp1	9.46	E3-ter
Retinol binding protein-3	180290	RBP3	10q11.2	Rbp3	14.14	B
Retinol binding protein-4	180250	RBP4	10q23-q24	Rbp4	19.19	D1-D
Usher type 1B	276900	MY7A	11q13.5	sh1	7.48	F1-F3
Waardenburg's syndrome type II	193510	MITF	3p14.1-p12.3	Mi	6.44	C2-C3
Waardenburg, PAX3-related	193500	PAX3	2q35	Pax3*	1.36	C4
Wolf–Hirschhorn	142983	MSX1	4p16.1	Msx1	5.14	B-E1

Mouse locus * refers to *in situ*. The figures in the mouse location column after the decimal point are estimated distances in centiMorgans (cM) from the centromere of the particular chromosome. Metabolic generalised deposition disorders (e.g. Tay–Sachs disease, Hurler's syndrome) have been omitted, as has colour blindness (a normal variant). MIM: Mendelian Inheritance in Man number (McKusick, 1988).
Data from Professor JH Edwards, personal communication as of December 1995.

particular species has a homologue in that same species, due to a duplication event at some point in its evolution, then these genes are *paralogous* to one another. A paralogy map may therefore be drawn by requesting an Oxford Grid of a species plotted against itself. This is particularly useful for species such as plants, where duplication events are more common.

Within the ACEDB comparative mapping database, double-clicking on a chosen Oxford Grid cell with the mouse brings up a text window with information on each of the homologous loci contained within it. Double-clicking on a locus name from this list will show its position on a chromosome map, should this be known. Simple statistics are available, such as a measure of the genome reorganisation that separates two species (Bengtsson *et al.*, 1993). The grid may be

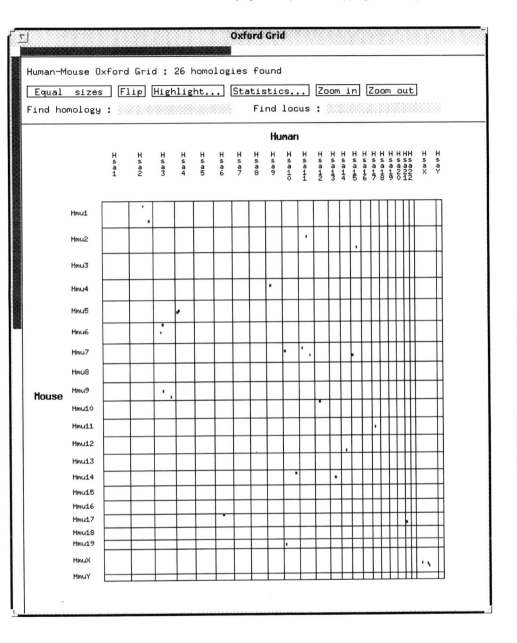

Fig. 17.3 An Oxford Grid of loci involved in eye disease

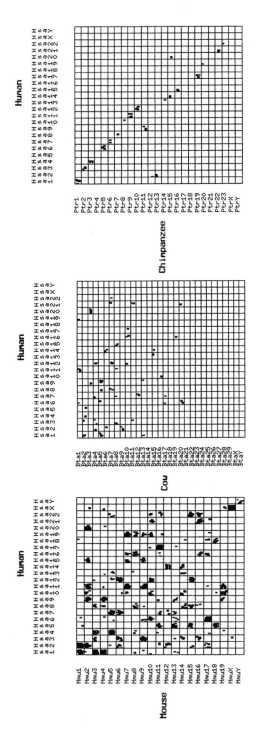

Fig 17.4 Three contrasting Oxford Grids

flipped, so that the two species exchange the axes on which they are drawn, and particular loci or homologies may be searched for. In addition to these functions the user is able, through the Oxford Grid, to examine the relationship between two chromosomes of different species, by means of the pairwise Chromosome Map. This provides the user with more detailed information about the nature and number of conserved segments within each cell.

The Pairwise Chromosome Map

The Pairwise Chromosome Map (PairMap) shows a single cell of an Oxford Grid in some detail. The PairMap in *Figure 17.5* shows homologous loci on human chromosome 14 and mouse chromosome 12. Homologies are shown on the screen as small blue boxes at their respective positions on the chromosomes of the two species, should these both be known. For example, looking at the enlarged 14,12 cell, the x-coordinate of each box is the locus' position on chromosome 14 in the human (or rather the midpoint of the range within which it is known to be located) and the y-coordinate is its position on chromosome 12 in the mouse. Homologies are shown with error bars relating to the known errors of their locations in each species. Well-mapped loci have small error bars and loci with less accurate locations will have large error bars.

The PairMap indicates segments that are conserved between the two species. These are easy to see as they are diagonal lines on the map. An inversion of a segment between the two chromosomes is easy to spot as a diagonal line going 'the wrong way'. A single click on a blue homology box causes the names of the two homologous loci to be displayed in the light blue header bar beneath the button menus and the box itself turns light green. Any homology may be selected by double-clicking on it with the mouse. Upon this a text window appears giving information on the homology and on homologies between the two loci involved and loci from other species. The PairMap has many other features, such as a search facility for loci and homologies, and an option to restrict the loci shown to a subset that is of interest to the user.

The One-to-Many Chromosome Map

A One-to-Many Chromosome Map (O2MMap) of human chromosome 14 with all mouse chromosomes is shown in *Figure 17.6*. This example shows that there are two or three likely conserved segments between human chromosome 14 and the

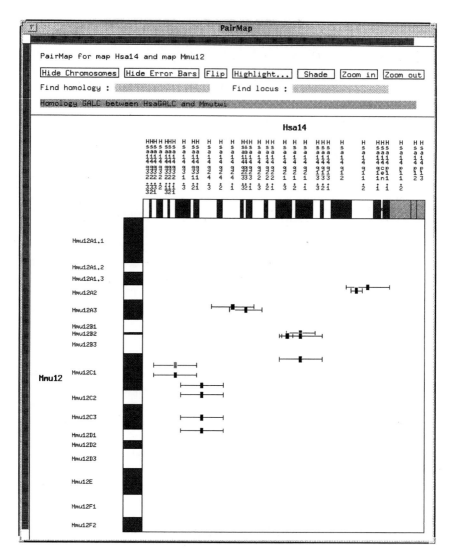

Fig. 17.5 A PairMap in the ACEDB Comparative Mapping Database

mouse, one with mouse chromosome 12 and one with mouse chromosome 14. Like the Pairwise Chromosome map, this map is accessed through the Oxford Grid. It effectively shows the loci in one of the Oxford Grid's rows or columns. If a column is selected, then an ideogram of the single chromosome of species 1 at the top of the column is placed alongside a grid showing all the chromosomes of species 2. Each

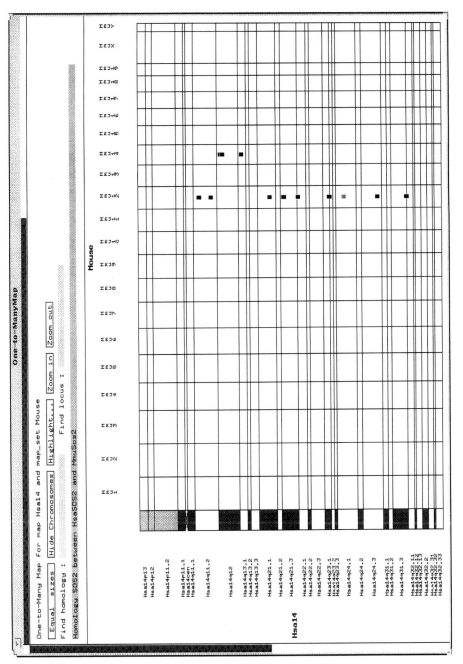

Figure 17.6 A O2MMap in the ACEDB Comparative Mapping Database

pair of homologous loci is placed in the grid, the y-coordinate of the homology box being at the midpoint of the range of the locus' location on the single chromosome of species 1 and the x-coordinate being in the centre of the column of its chromosome in species 2. If an Oxford Grid row is selected, then a single chromosome of species 2 is gridded with all chromosomes of species 1 in a similar way.

A O2MMap shows the distribution of loci on the single chromosome and their homologues throughout the chromosomes of the other species. Conserved segments between the two species become apparent as a clustered vertical line of blue boxes. It also highlights duplicated segments in species such as plants as the user will see two vertical clusters of blue boxes at equal heights on the map. A single click on a blue box causes the names of the two homologous loci to be displayed in the light blue header bar beneath the button menus, with the clicked box turning light green. Any homology may be selected by double-clicking on it with the mouse. Upon this a text window appears giving information on the homology and on homologies of the two loci with loci from other species. The O2MMap also has additional features similar to those of the PairMap.

The Species Grid

A Species Grid showing homologues of loci on human chromosome 14 is shown in *Figure 17.7*. Such a grid shows all loci on a particular chromosome that have homologues in other species. The y-coordinate of the homologue is the locus' position on the single chromosome. The x-coordinate is the column midpoint of the second species, and each homology is shown with the chromosome number in the second species to its right. Conserved segments between the single chromosome and the other species may be inferred from this. Like the O2MMap, these will be apparent as clustered vertical lines of loci. A single click on a blue box causes the names of the two homologous loci to be displayed in the light blue header bar beneath the title and the box itself turns light green. Any homology may be selected by double-clicking on it with the mouse. Upon this a text window appears giving information on the homology and on homologies of the two loci with loci from other species. The Species Grid has additional features similar to those of the PairMap.

Conclusion

We have seen how comparative mapping diagrams can aid the mapping of genes by interspecies analyses of gene order. The ACEDB comparative mapping

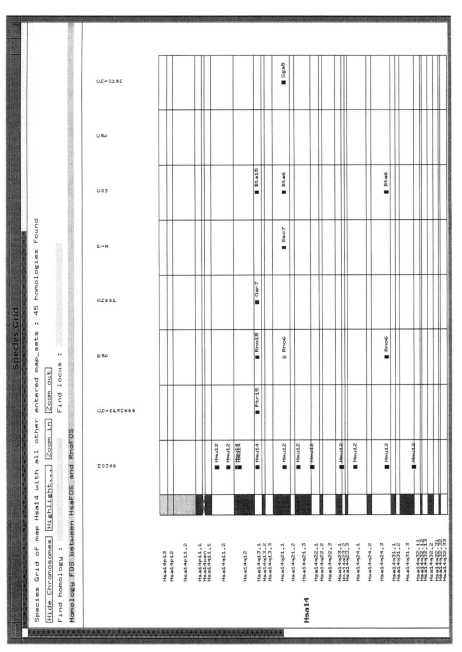

Figure 17.7 A Species Grid in the ACEDB Comparative Mapping Database

database provides an automated procedure to speed up such analyses. A database, ZOODB, which uses this software is available to registered users of the UK MRC Human Genome Mapping Project Resource Centre at Hinxton (see 'http://www.mrc.ac.uk/'). The ACEDB comparative mapping software may be obtained by requesting it directly from the author. It will be publicly available shortly, with the official ACEDB software release.

Acknowledgements

This project has been funded by the Medical Research Council's UK Human Genome Mapping Project under the supervision of Professor J. H. Edwards of the Department of Biochemistry, Oxford. Dr A. G. Searle of the MRC Mammalian Genetics Unit in Harwell, Oxon has provided the human–mouse homology data. Much of this work has been carried out in collaboration with Michelle Kirby, also of this unit. I must also thank Dr Otto Ritter of *Deutsches Krebsforschungs-zentrum* (DKFZ), Heidelberg for his help during this project, Dr Sam Cartinhour of Texas A & M University for his many helpful comments and ideas on how to improve the maps, and Dr Dave Matthews of Cornell University for his help with the comparative mapping of plant species.

References

Bengtsson, B.O., Klinga Levan, K., Levan, G. Measuring genome reorganization from synteny data. (1993) *Cytogenet. Cell Genet.* **64 (3–4):** 198–200.

Buckle, V.J., Edwards, J.H., Evans, E.P., Jonasson, J.A., Lyon, M.F., Peters, J., Searle, A.G., Wedd, N.S. Chromosome maps of man and mouse II. (1984) *Clin. Genet.* **26 (1):** 1–11.

Dalton, T.P., Edwards, J.H., Evans, E.P., Lyon, M.F., Parkinson, S.P., Peters, J., Searle, A.G. Chromosome maps of man and mouse. (1981) *Clin. Genet.* **20 (6):** 407–415.

Durbin, R., Thierry-Mieg, J. (1991) A *C. elegans* DataBase. Documentation, code and data available from anonymous FTP servers at lirmm.lirmm.fr, cele.mrc-lmb.cam.ac.uk and ncbi.nlm.nih.gov

Edwards, J.H. The Oxford Grid. (1991) *Ann. Hum. Genet.* **55 (1):** 17–31.

Edwards, J.H. Comparative genome mapping in mammals. (1994) *Curr. Opin. Genet. Dev.* **4 (6):** 861–867.

McKusick, V.A. *Mendelian Inheritance in Man.* (1988) Johns Hopkins University Press, Baltimore.

O'Brien, S.J. Mammalian genome mapping: lessons and prospects. (1991) *Curr. Opin. Genet. Dev.* **1 (1):** 105–111.

Searle, A.G., Peters, J., Lyon, M.F., Evans, E.P., Edwards, J.H., Buckle, V.J. Chromosome maps of man and mouse III. (1987) *Genomics* **1 (1):** 3–18.

Searle, A.G., Peters, J., Lyon, M.F., Hall, J.G., Evans, E.P., Edwards, J.H., Buckle, V.J. Chromosome maps of man and mouse IV. (1989) *Ann. Hum. Genet.* **53 (2):** 89–140.

SECTION V

APPLICATIONS

Chapter 18
From gene mapping to the identification of mutations: the example of the Hirschsprung's disease genes

Marcella Devoto

◆

Introduction

Hirschsprung's disease (HSCR), often also called aganglionic megacolon, is a congenital disorder characterized by the absence of the intrinsic ganglion cells of the submucosal and myenteric plexuses of the hindgut. This results in symptoms varying from severe functional intestinal obstruction and megacolon to milder constipation, which is often the only symptom in older patients. For the more severe cases, the only possible treatment is surgery and removal of the abnormal bowel (Rudolph and Benaroch, 1995).

HSCR is a neurocristopathy or a disorder of the neural crest and its derivatives, possibly caused by a premature arrest of the migration of neural crest cells towards the anal end of the rectum or by a defect in colonization and differentiation of the same cells. The earlier the migration stops, the larger the aganglionic segment. Accordingly, HSCR cases are generally classified into short and long forms, depending upon whether the aganglionic tract extends from the rectum up to the splenic flexure, which is the more frequent form (accounting for about 80% of cases) and is the so-called short form, or also includes the transverse colon and the proximal part of the intestine (i.e. the long form, or total aganglionosis). It is interesting to note that there is a difference in embryonic origin between the

GENETIC MAPPING OF DISEASE GENES
ISBN 0-12-232735-7

gastrointestinal tract that includes the duodenum and up to approximately two-thirds of the transverse colon, which originates from the embryonic midgut, and the remaining part of the colon and the rectum, which originate from the embryonic hindgut (Rudolph and Benaroch, 1995).

The genetics of HSCR

Several lines of evidence favor the hypothesis of a genetic component in HSCR etiology (Passarge, 1967). First, the increased risk to relatives (and to sibs in particular) with respect to a general population incidence of 0.02%. Large pedigrees with recurrence of HSCR disease have often been observed. Second, the frequent association of HSCR with other malformations and with chromosomal abnormalities, including Waardenburg's syndrome, Smith–Lemli–Opitz syndrome, trisomy 21, partial trisomy of chromosomes 22 and 11, and deletions, in particular of chromosomes 10 and 13. Finally, the existence of several animal models of aganglionic megacolon where the disease follows a simple monogenic form of inheritance. However, in general HSCR is considered to be a complex disease with multifactorial inheritance with multiple genes and the environment playing a role in determining the clinical phenotype.

The first important step in understanding the genetics of HSCR came from segregation analysis carried out by Badner et al. (1990). This study confirmed previously reported observations that showed a sex ratio of approximately 4 : 1 males to females, and an overall risk of approximately 4% in sibs compared to 0.02% in the general population. In addition, it was clearly shown that the sex ratio decreased and the risk to sibs increased with increasing extent of the aganglionosis, suggesting that a different etiology might underlie the different clinical forms.

In accordance with these observations, segregation analysis was performed separately for the patients with long and short forms of aganglionosis. Results showed that both a dominant and an additive model fitted the data for patients with long-segment HSCR, but the sporadic, multifactorial, and recessive models could all be rejected. Similar results were obtained for the short-segment HSCR families. However, when these were further subdivided into colonic segment and rectosigmoid segment forms, in accordance with recurrence risk calculations that showed a correlation with the extent of the aganglionosis, it appeared that the recessive and the additive model fitted the data for the shortest form equally well, but the dominant model could be rejected (Badner et al., 1990).

240

Genetic mapping of HSCR

Segregation analysis set the basis for a formal genetic model of HSCR and several groups then started carrying out linkage analysis for the familial cases. They used markers from the candidate regions identified by the associated chromosomal abnormalities, and in particular from chromosomes 11, 13, 21, and 22. However, no significant results were obtained until a *de novo* interstitial deletion of the proximal long arm of chromosome 10 was observed in a young patient with total aganglionosis in 1992 (Martucciello *et al.*, 1992). Following isolation of the two chromosomes 10 from this patient in two different somatic cell hybrids, a set of markers was identified that mapped inside the deletion. Linkage analysis carried out in a sample of 15 HSCR families using the same markers clearly showed positive significant results with a maximum lod-score of 4.26 at a recombination fraction of 3.7% from marker D10S196 (Lyonnet *et al.*, 1993).

When the same families were divided into long and short forms, significant results were only obtained for the long-form families, although a formal test of heterogeneity was not significant and therefore the possibility of genetic homogeneity of the two forms could not be excluded at that time (Lyonnet *et al.*, 1993). Totally equivalent results were obtained at the same time in an independent study carried out using a different panel of families and different markers from the same region, in which evidence for linkage to chromosome 10q11.2 and suggestive, but not conclusive, indication of heterogeneity, were found (Angrist *et al.*, 1993).

The *ret* protooncogene

The region of chromosome 10 identified above was about 10 cM long and was therefore likely to contain many genes, although the only gene known to map inside that region at that time was the *ret* protooncogene. The same region was also known to contain the locus of three other neurocristopathies, namely multiple endocrine neoplasia type 2A and 2B (MEN2A and MEN2B), and medullary thyroid carcinoma (MTC). At about the same time that linkage of HSCR to chromosome 10 was found, mutations of the *ret* protooncogene were found to be responsible for cases of MEN2A, MEN2B, and MTC (Smith *et al.*, 1994).

The *ret* protooncogene itself had not been tested for linkage to HSCR in the initial chromosome 10 screening. However, it soon became evident from the observation of other deletions that further restricted the candidate region for HSCR, that *ret* itself was a strong candidate. Linkage analysis carried out using a marker located near the 3′ end of *ret* produced a significant result at a recombination fraction of 0. In addition, a family was identified in which a microdeletion of *ret* itself segregated with the disease in all patients and asymptomatic carriers of the disease (Luo *et al.*, 1993).

The next step was clearly to test other HSCR patients for mutations in the *ret* gene. In this way, a variety of different mutations scattered along the whole coding region of *ret* were soon found to be responsible for both sporadic and familial cases of HSCR (Edery *et al.*, 1994; Romeo *et al.*, 1994). In contrast, mutations in the *ret* protooncogene causing MEN2A and familial MTC are almost all concentrated in the cysteine residues in exons 10, 11, 13, and 14, and a single mutation affecting methionine 918 has been found to be responsible for almost all cases of MEN2B.

Although MEN2A, MEN2B, MTC, and HSCR are all disorders of the neural crest and its derivatives, they differ from each other in a variety of ways. For example, MEN2B is associated with hyperplasia and dysplasia of the enteric ganglia, which are absent in HSCR cases, and only very few cases of an association between MEN2A and HSCR have been reported. In addition to *ret*, mutations in several other oncogenes and tumor suppressor genes are known to be associated with developmental disorders, as in the case of *pax*3, responsible for rhabdomyosarcoma and Waardenburg's syndrome, and of the von Hippel–Lindau gene, which is associated with renal cell carcinoma. This phenomenon, known as phenotypic diversity due to allelic series, can in fact be extended to over 100 genes whose mutations may cause two or more clinically different, apparently unrelated, disorders (Romeo and McKusick, 1994).

The search for other HSCR genes

Mutation analysis with the *ret* protooncogene soon demonstrated that defects in this gene only accounted for a relatively small proportion of HSCR cases. In fact, the detection rate was always relatively low (around 20–25%), and it was higher in the long-form families than in the short-form families, confirming the hypothesis of etiologic heterogeneity of the two forms suggested by the genetic analysis (Luo *et al.*, 1994).

A major breakthrough in the search for the other HSCR genes came from the study of a large Mennonite kindred with a very high incidence of HSCR (Puffenberger *et al.*, 1994b). A total of 61 nuclear pedigrees were observed in which all parents of affected individuals were related with a degree of consanguinity close to that of second or third cousins, and all individuals descended from the same ancestral couple. This finding led to the hypothesis of a recessive transmission of HSCR in this kindred, and this was further supported by the fact that only in a few cases were one of the patients' parents also affected, an observation which is compatible with the presence of a relatively frequent recessive allele. Based on the hypothesis that all patients should share the same disease allele and therefore also the same alleles at the closely linked markers, a search for regions in the genome shared identical by descent (IBD) was carried out in only three of the 61 nuclear pedigrees. After identifying a few candidate regions the analysis was extended to all the available nuclear pedigrees by testing for linkage disequilibrium, comparing the marker allele frequencies among transmitted and non-transmitted chromosomes.

In this way, a region spanning nine markers on chromosome 13q22 was identified at which a common haplotype was present in most of the disease chromosomes, but was very rare in the non-transmitted chromosomes (Puffenberger *et al.*, 1994b).

The rodent models and the endothelin receptor type B gene

The mapping of a HSCR gene on chromosome 13 was consistent with two other findings. First, it was known that a number of patients carried deletions of chromosome 13q. Second, one of the mouse models of HSCR, the recessive piebald-lethal, had been mapped to mouse chromosome 14 in a region homologous to human chromosome 13. These mutant mice are characterized by an almost completely white coat and by megacolon. A milder allelic form, known as piebald, manifests white spotting in only about 20% of the coat and almost never has megacolon.

The association of a pigmentary defect with aganglionic megacolon is not limited to the piebald mutants. The same association is found in mice with lethal-spotting (ls) and dominant megacolon (Dom), and in the rat with spotting lethal (sl). In humans, Waardenburg's syndrome, which is characterized by hearing loss and pigmentary defects, is often associated with HSCR. This association reflects the common neural crest origin of the epidermal melanocytes and the myenteric ganglion neurons, which apparently share a common pathway in a crucial phase of their development.

Hosoda *et al.* (1994) were able to show that a null mutation created by the targeted disruption of the endothelin receptor type B (ENDRB) gene produced a phenotype characterized by white spotting and megacolon identical to that of the piebald-lethal mutants. Further experiments showed that the ENDRB null mutation and piebald-lethal were allelic, and that the piebald-lethal phenotype was in fact due to a deletion of the whole ENDRB gene. More recently, a deletion in the

243

ENDRB gene was also found to be responsible for the rat megacolon known as spotting lethal (Ceccherini *et al.*, 1995). The gene responsible for another mouse megacolon, Dom, has also been mapped on mouse chromosome 15 in a region homologous to human chromosome 22, which has been associated with partial trisomy in HSCR patients (Puliti *et al.*, 1995).

The human ENDRB gene was located in the same region of chromosome 13, which showed linkage disequilibrium with HSCR in the Mennonite kindred. A search for mutations of this gene among these patients led to the identification of a mutation in exon 4, which results in the substitution of a cysteine for trypto-phan (W276C) (Puffenberger *et al.*, 1994a). Haplotype analysis revealed that 40 of 56 haplotypes associated with HSCR in the Mennonite kindred also carried the W276C mutation. The HSCR chromosomes that did not carry W276C also carried a different haplotype at the neighboring markers, and at least five individuals with clear signs of HSCR did not carry the W276C mutation.

It became clear, therefore, that even in this relatively homogeneous and highly inbred population different genes were causing HSCR disease. A role of *ret* has been suggested by linkage disequilibrium analysis in the same Mennonite kindred, but no mutations have so far been found in the *ret* gene.

Analysis of the genotypes of 150 patients, their parents, and sibs, showed that the penetrance of W276C was less that 100% even in the homozygous state, and it was higher in males than in females, in accordance with the previous obser-vation of a sex difference in penetrance (Puffenberger *et al.*, 1994a).

Genotype and phenotype correlation analysis showed that the W276C mutation was often associated with pigmentation defects and deafness, which are characteristic of Waardenburg's syndrome. Following this observation, patients with Waardenburg's syndrome who were negative for mutations in *pax*3 or microphthalmia-associated transcription factor (MITF) were investigated for mutations in ENDRB. At least one other case is known of a family with Waardenburg's syndrome–HSCR and a mutation in the ENDRB gene (Attie *et al.*, 1995). Few other HSCR patients unrelated to the Mennonite kindred have been found to carry mutations in the ENDRB gene (Kusafuka *et al.*, 1996). Two mutations in the endothelin 3 itself, which is mutated in the mouse lethal-spotted, have also been found in Waardenburg's syndrome–HSCR patients, but do not seem to be very frequent (Edery *et al.*, 1996; Hofstra *et al.*, 1996). More recently, Auricchio *et al.* (1996) have identified a locus on the X chromosome responsible for a form of intestinal pseudoobstruction, which in view of the sex-dependent penetrance of the other HSCR genes may become an interesting candidate in the search for modifier genes.

Although not concluded, the search for genes implicated in HSCR shows how the genetic basis of a complex multifactorial disease can be unravelled by combining different genetic approaches.

References

Angrist, M., Kauffman, E., Slaugenhaupt, S.A., Matise, T.C., Puffenberger, E.G., Washington, S.S., Lipson, A., Cass, D.T., Reyna, T., Weeks, D.E., Sieber, W., Chakravarti, A. A gene for Hirschsprung disease (megacolon) in the pericentromeric region of human chromosome 10. (1993) *Nat. Genet.* **4 (4):** 351–356.

Attie, T., Till, M., Pelet, A., Amiel, J., Edery, P., Boutrand, L., Munnich, A., Lyonnet, S. Mutation of the endothelium–receptor B gene in Waardenburg–Hirschsprung disease. (1995) *Hum. Mol. Genet.* **4:** 2407–2409.

Auricchio, A., Brancolini, V., Casari, G., Milla, P.J., Smith, V.V., Devoto, M., Ballabio, A. Locus for a novel syndromic form of neuronal intestinal pseudoobstruction maps to Xq28. (1996) *Am. J. Hum. Genet.* **58 (4):** 743–748.

Badner, J.A., Sieber, W.K., Garver, K.L., Chakravarti, A. A genetic study of Hirschsprung disease. (1990) *Am. J. Hum. Genet.* **46 (3):** 568–580.

Ceccherini, I., Zhang, A.L., Matera, I., Yang, G., Devoto, M., Romeo, G., Cass, D.T. Interstitial deletion of the endothelin-B receptor gene in the spotting lethal (sl) rat. (1995) *Hum. Mol. Genet.* **4 (11):** 2089–2096.

Edery, P., Lyonnet, S., Mulligan, L.M., Pelet, A., Dow, E., Abel, L., Holder, S., Nihoul Fekete, C., Ponder, B.A., Munnich, A. Mutations of the RET proto-oncogene in Hirschsprung's disease. (1994) *Nature* **367 (6461):** 378–380.

Edery, P., Attie, T., Amiel, J., Pelet, A., Eng, C., Hofstra, R.M.W., Martelli, H., Bidaud, C., Munnich, A., Lyonnet, S. Mutation of the endothelin 3 gene in the Waardenburg Hirschsprung disease (Shah Waardenburg syndrome). (1996) *Nat. Genet.* **12 (4):** 442–444.

Hofstra, R.M.W., Osinga, J., Tansindhunata, G., Wu, Y., Kamsteeg, E.J., Stulp, R.P., Vanravenswaaijarts, C., Majoorkrakauer, D., Angrist, M., Chakravarti, A., Meijers, C., Buys, C.H.C.M. A homozygous mutation in the endothelin 3 gene associated with a combined Waardenburg type 2 and Hirschsprung phenotype (Shah Waardenburg syndrome). (1996) *Nat. Genet.* **12 (4):** 445–447.

Hosoda, K., Hammer, R.E., Richardson, J.A., Baynash, A.G., Cheung, J.C., Giaid, A., Yanagisawa, M. Targeted and natural (piebald-lethal) mutations of endothelin-B receptor gene produce megacolon associated with spotted coat color in mice. (1994) *Cell* **79 (7):** 1267–1276.

Kusafuka, T., Wang, Y.P., Puri, P. Novel mutations of the endothelin B receptor gene in isolated patients with Hirschsprung's disease. (1996) *Hum. Mol. Genet.* **5 (3):** 347–349.

Luo, Y., Ceccherini, I., Pasini, B., Matera, I., Bicocchi, M.P., Barone, V., Bocciardi, R., Kaariainen, H., Weber, D., Devoto, M., Romeo, G. Close linkage with the RET protooncogene and boundaries of deletion mutations in autosomal dominant Hirschsprung disease. (1993) *Hum. Mol. Genet.* **2 (11):** 1803–1808.

Luo, Y., Barone, V., Seri, M., Bolino, A., Bocciardi, R., Ceccherini, I., Pasini, B., Tocco, T., Lerone, M., Cywes, S., Moore, S., Vanderwinder, J.M., Abramowicz, M.J., Kristoffersson, U., Larsson, L.T., Hamel, B.C.J., Silengo, M., Martucciello, G., Romeo, G. Heterogeneity and low detection rate of RET mutations in Hirschsprung disease. (1994) *Eur. J. Hum. Genet.* **2 (4):** 272–280.

Lyonnet, S., Bolino, A., Pelet, A., Abel, L., Nihoul Fekete, C., Briard, M.L., Mok Siu, V., Kaariainen, H., Martucciello, G., Lerone, M., Puliti, A., Luo, Y., Weissenbach, J., Devoto, M., Munnich, A., Romeo, G. A gene for Hirschsprung disease maps to the proximal long arm of chromosome 10. (1993) *Nat. Genet.* **4 (4):** 346–350.

Martucciello, G., Biccochi, M.P., Dodero, P., Lerone, M., Silengo, M.C., Puliti, A., Gimelli, G., Romeo, G., Jasonni, V. Total colonic aganglionosis associated with interstitial deletion of the long arm of chromosome 10. (1992) *Pediatr. Rev. Int.* **7**: 308–310.

Passarge, E. The genetics of Hirschsprung's disease. Evidence for heterogeneous etiology and a study of sixty-three families. (1967) *N. Engl. J. Med.* **276 (3)**: 138–143.

Puffenberger, E.G., Hosoda, K., Washington, S.S., Nakao, K., deWit, D., Yanagisawa, M., Chakravarti, A. A missense mutation of the endothelin-B receptor gene in multigenic Hirschsprung's disease. (1994a) *Cell* **79 (7)**: 1257–1266.

Puffenberger, E.G., Kauffman, E.R., Bolk, S., Matise, T.C., Washington, S.S., Angrist, M., Weissenbach, J., Garver, K.L., Mascari, M., Ladda, R., Slangenhaupt, S.A., Chakravarti, A. Identity-by-descent and association mapping of a recessive gene for Hirschsprung disease on human chromosome 13q22. (1994b) *Hum. Mol. Genet.* **3 (8)**: 1217–1225.

Puliti, A., Prehu, M.O., Simonchazottes, D., Ferkdadji, L., Peuchmaur, M., Goossens, M., Guenet, J.L. A high resolution genetic map of mouse chromosome 15 encompassing the dominant megacolon (Dom) locus. (1995) *Mam. Genome* **6 (11)**: 763–768.

Romeo, G., McKusick, V.A. Phenotypic diversity, allelic series and modifier genes. (1994) *Nat. Genet.* **7 (4)**: 451–453.

Romeo, G., Ronchetto, P., Luo, Y., Barone, V., Seri, M., Ceccherini, I., Pasini, B., Bocciardi, R., Lerone, M., Kaariainen, H., Martucciello, G. Point mutations affecting the tyrosine kinase domain of the RET proto-oncogene in Hirschsprung's disease. (1994) *Nature* **367 (6461)**: 377–378.

Rudolph, C., Benaroch, L. Hirschsprung disease. (1995) *Pediatr. Rev. Int.* **16 (1)**: 5–11.

Smith, D.P., Eng, C., Ponder, B.A. Mutations of the RET proto-oncogene in the multiple endocrine neoplasia type 2 syndromes and Hirschsprung disease. (1994) *J. Cell Sci.* **18 (Suppl.)**: 43–49.

Chapter 19
Retinal dystrophies: a molecular genetic approach

K. Evans and Shomi S. Bhattacharya

Understanding of the molecular genetics underlying retinal dystrophies has increased enormously over the last decade. This has been a natural consequence of progress in molecular biology and recombinant DNA technology. The development of genetic markers based upon the wealth of DNA polymorphisms identified throughout the genome, and the application of this information in linkage analysis has been especially important. The best example to illustrate this progress is the work that has been undertaken on the molecular genetics of autosomal dominant retinitis pigmentosa (RP).

Retinitis pigmentosa

RP identifies a group of disorders that is associated with night blindness (nyctalopia) and constricted peripheral visual field as early manifestations. Major differences are found in the age of onset of symptoms, the nature of visual loss, and speed of progression between the different disorders within this group. Abnormal, bilateral, symmetrical 'bone spicule-like' intraretinal pigmentation is seen, which can later develop into extensive chorioretinal atrophy with attenuation of the retinal vasculature and secondary optic atrophy. In such advanced stages, visual acuity is reduced, often leading to blindness (Humphries *et al.*, 1994).

Epidemiological studies of RP around the world consistently report a frequency of approximately 1 in 5,000 of the general population (Grondahl, 1987).

GENETIC MAPPING OF DISEASE GENES
ISBN 0-12-232735-7

There are approximately 100,000 people with RP in Europe and a similar number in the USA. A family history of RP can be demonstrated in approximately 50% of new cases, and all three mendelian inheritance patterns (autosomal dominant, recessive and X-linked pedigrees) exist. In 15–50% of cases no inheritance pattern is clearly identified (simplex RP). Autosomal dominant inheritance accounts for approximately 17–25% of all cases (Bundey and Crews, 1984).

Linkage analysis

Linkage analysis is a powerful technique for localising genetic abnormalities in the genome. The method is based on the assumption that if two genetic loci are close together in the genome they will tend to be inherited together through successive meiosis. Everyone carries two alleles at each genetic locus and with a probability of 0.5 will transmit a particular one of the two to an offspring. If two gene loci are considered, there are four possible selections of two non-allelic genes, one from each of the two loci, which are transmitted in an approximate 1:1:1:1 ratio. If for specific pairs of loci, a deviation from this ratio is observed in the sense that the two non-allelic genes received from one parent of an individual are transmitted together to the offspring of this individual then they are genetically linked and close together in the genome (Ott, 1986).

Usually, genetic linkage is not complete so that even for such 'linked loci', non-allelic genes received from two parents recombine and an individual may transmit to an offspring one gene at one locus received from the mother and one gene at another linked locus received from the father. The proportion of such recombinations out of all opportunities to recombine is called the recombination fraction. For unlinked gene loci this fraction is 0.5, for linked loci it is less. The recombination fraction can be looked upon as a measure of genetic distance between loci and is a central calculation in the logarithm of odds (lod) score method of linkage analysis. Linkage between two loci is significant when the lod score is ≥3; this corresponds to the odds of linkage of being 1,000:1. A lod score value of ≤ –2 is considered to indicate significant exclusion of linkage at that recombination value (Ott, 1986).

Despite assumptions and simplifications the calculations in the lod score method, especially multipoint calculations are monumental. To reduce this and improve accuracy, a variety of computer software packages have become available containing relevant programs. LINKSYS (Attwood and Bryant, 1988) is a data management package, a database, specifically designed to convert data into a form suitable for linkage calculation programs. Family files are generated giving pedigree and disease status information. Locuslib is a locus library, storing data on disease and marker loci, while phenolib is a phenotype library containing details on all possible phenotypes for each locus stored in locuslib. LINKAGE (Lathrop *et al.*, 1984) is a suite of programs: MLINK calculates likelihoods for given parameter values, ILINK is an iterative version of MLINK for estimating maximum likelihood estimates of parameters, and LINKMAP calculates likelihoods for the localisation of one locus of unknown position relative to a known map of several marker loci. Despite advances in computer technology, memory limits are often exceeded. Desktop personal computers then need to be replaced with larger mainframe machines such as those available at the Human Genome Mapping Project Resource Centre (HGMP-RC, Cambridge) accessible via telnet. Significantly more memory can be accessed using operating systems such as UNIX.

In linkage analysis the unknown human disease gene locus is compared with a marker locus based upon the wealth of polymorphisms known to exist throughout the genome. Two types of DNA polymorphism are seen. The first is based on single nucleotide base changes at specific loci. These create or destroy restriction sites, which can be recognised and cut by specific restriction endonucleases giving fragments of variable size, which are usually defined as restriction fragment length polymorphisms (RFLPs). Markers that recognise such RFLPs are termed polymorphic and are therefore valuable for genetic linkage studies in families. Generally RFLP-based markers have only two alleles corresponding to the presence or absence of the restriction site (Davies and Read, 1992) and are therefore often uninformative in linkage analysis. The second type of marker identifies simple runs of tandemly repeated sequence that can be detected using the polymerase chain reaction (PCR). The repeating unit may be ten or so nucleotides long (minisatellites) (Jeffreys *et al.*, 1985) or only two or three nucleotides (microsatellites) (Weber and May, 1989). Occasionally runs of a single nucleotide are polymorphic (Kumar-Singh *et al.*, 1991).

Using the mathematics of linkage analysis, highly polymorphic DNA markers and the laboratory techniques such as the PCR, linkage between disease loci and marker loci can be sought in large pedigrees. DNA samples can be obtained from peripheral blood samples of individual family members whose disease status and position in the pedigree is known. Using these methods, a total of nine autosomal dominant RP loci has been found (*Table 19.1*).

Physical mapping strategies in RP

Once sufficiently small genomic regions containing dominant RP loci have been identified using linkage analysis, studies have proceeded to the construction of

Table 19.1. Chromosomal localisations of autosomal dominant RP

Chromosomal location	Gene	Reference
1p13-q23	–	Xu *et al.*, 1996
3q21	Rhodopsin	McWilliam *et al.*, 1989
6p	Peripherin/Retinal degeneration slow (RDS)	Farrar *et al.*, 1991
7p14	–	Inglehearn *et al.*, 1993
7q	–	Jordan *et al.*, 1993
8cen	–	Blanton *et al.*, 1991
17p13.1	–	Greenberg *et al.*, 1994
17q	–	Bardien *et al.*, 1995
19q13.4	–	Al-Maghtheh *et al.*, 1994

physical contigs of the regions of interest. Libraries of cloned genomic DNA fragments of various sizes are available. These fragments have been amplified within various vectors, for example yeast to produce yeast artificial chromosomes (YACs). YACs mapping to the region of interest are isolated from libraries by probing with microsatellite markers linked to the disease locus. An overlapping ordered arrangement of clones can then be constructed by sequencing the terminal ends of each fragment and using this sequence to 'pull out' the next clone from the library. Such methodology has been used to construct a physical map of the 4.8 Mb region containing the dominant RP locus RP9 on chromosome 7p (Keen *et al.*, 1995). Various methods such as 'CpG island' mapping (Cotton, 1993) and cDNA selection (Parimoo *et al.*, 1993) are now being used to identify the coding sequence of candidate genes within the region.

Rhodopsin and peripherin/RDS mutation screening in RP

Many techniques are available for the identification of disease-causing mutations. Single strand conformation polymorphism (SSCP) analysis (Orita *et al.*, 1989), denaturing gradient gel electrophoresis (DGGE) (Nichols *et al.*, 1993), and heteroduplex analysis (Keen *et al.*, 1994) have been used to identify mutations in retinal dystrophies. When a specific mutation is to be sought, allele-specific oligonucleotides (ASOs) have been used (Inglehearn *et al.*, 1991). DNA primers are designed that recognise the mutant copy. In this technique DNA is usually immobilised on filters in the form of 'dot blots'. Normal DNA will only hybridise to ASOs that recognise normal sequence, whereas DNA from individuals carrying the mutation will hybridise to normal and mutation-specific ASOs because the sample will contain two copies, one normal and one mutant.

Genetic mutations in only two genes have so far been associated with dominant RP. Rhodopsin encodes the membrane-bound photopigment found in all rod photoreceptors. Mapping to the same chromosome 3q region implicated in dominant RP by McWilliam *et al.* (1989) lead to the identification of a codon 23 mutation (Dryja *et al.*, 1990) accounting for 15% of all dominant RP cases in the USA. This particular mutation has not yet been found outside the USA. To date over 60 different mutations in the rhodopsin gene have been causally associated with dominant RP (Al-Maghtheh *et al.*, 1993). Peripherin/RDS maps to chromosome 6p12 and encodes a membrane-bound protein important for the structural integrity of both rod and cone photoreceptor outer segments. Mutation screening has identified 18 peripherin/RDS mutations in different dominant RP families (Farrar *et al.*, 1994). As would be expected peripherin/RDS mutations have also been seen in pedigrees expressing principally cone-defective retinal dystrophies (Wells *et al.*, 1993).

Molecular genetics as a basis for phenotypic subclassification

Dominant RP has been subclassified into type I (D) and type II (R) forms. The former is characterised as 'diffuse' in that widespread loss of rod function is associated with relative preservation of cone function at some stage of the disease. The latter is characterised by 'regional' distribution of disease with loss of both rod and cone function in some areas, but near normal function in others. This subclassification to some extent identifies genetic subtypes. Mainly type I functional loss has been associated with rhodopsin mutations (Inglehearn *et al.*, 1992) with disease severity linked to age. Type II functional loss has been described with peripherin/RDS mutations, again with severity consistently related to age. Detailed phenotype descriptions for pedigrees linked to chromosome 1, 7q, 8cen, 17p and 17q are as yet unpublished. A phenotype with variable expression (i.e. disease severity not linked to age and an R type pattern of deficit) has been described for the dominant RP pedigree linked to chromosome 7p. R type functional deficit has also been seen in families linked to the 19q locus. Severity of disease has been described as bimodal with affected family members being either severely affected from the second decade of life or asymptomatic with normal clinical examinations as late as the eighth decade of life (Evans *et al.*, 1995). In smaller, less well-documented pedigrees, bimodal expressivity may seem to present as cases of simplex RP. This, together with the fact that five pedigrees with this novel phenotype have now been linked to chromosome 19, suggests that the 19q locus may make a relatively large contribution to the aetiology of dominant

RP (Al-Maghtheh *et al.*, 1996). To a limited extent therefore, identification of the genetic abnormality can be used to predict the type of visual loss and the variation in severity that might be expected in the pedigree under study.

Application of molecular genetics in clinical practice

The genetic counselling of families with retinal dystrophy involves refined clinical diagnosis, an assessment of risk to subsequent children, and an appraisal of therapeutic options. The genetic techniques described above are now sufficiently well established to allow for the widespread use of molecular genetic assessment as an aid to diagnosis. In addition, advances in understanding of the molecular genetic abnormalities underlying retinal dystrophies would also seem to offer the best hope for developing effective treatment strategies. Finally the efficacies of different therapeutic modalities may be better assessed if they are evaluated on genetically homogenous groups of retinal dystrophies.

References

Al-Maghtheh, M., Gregory, C., Inglehearn, C., Hardcastle, A., Bhattacharya, S. Rhodopsin mutations in autosomal dominant retinitis pigmentosa. (1993) *Hum. Mutat.* **2:** 249–255.

Al-Maghtheh, M., Inglehearn, C.F., Keen, T.J., Evans, K., Moore, A.T., Jay, M., Bird, A.C., Bhattacharya, S.S. Identification of a sixth locus for autosomal dominant retinitis pigmentosa on chromosome 19. (1994) *Hum. Mol. Genet.* **3:** 351–354.

Al-Maghtheh, M., Vithana, E., Tarttelin, E., Jay, M., Evans, K., Moore, A.T., Bhattacharya, S.S., Inglehearn, C.F. Evidence for a major retinitis pigmentosa locus on 19q13.4 (RP11) and association with a unique bimodal expressivity phenotype. (1996) *Am. J. Hum. Genet.* **59:** 864–871.

Attwood, J., Bryant, S. A computer program to make linkage analysis with LIPED and LINKAGE easier to perform and less prone to input errors. (1988) *Ann. Hum. Genet.* **52 (3):** 259.

Bardien, S., Ebenezer, N., Greenberg, J., Inglehearn, C.F., Bartmann, L., Goliath, R., Beighton, P., Ramesar, R., Bhattacharya, S.S. An eighth locus for autosomal dominant retinitis pigmentosa is linked to chromosome 17q. (1995) *Hum. Mol. Genet.* **8:** 1459–1462.

Blanton, S.H., Heckenlively, J.R., Cottingham, A.W., Friedman, J., Sadler, L.A., Wagner, M., Friedman, L.H., Daiger, S.P. Linkage mapping of autosomal dominant retinitis-pigmentosa (RP1) to the pericentric region of human chromosome-8. (1991) *Genomics.* **11:** 857–869.

Bundey, S., Crews, S.J. A study of retinitis pigmentosa in the City of Birmingham. II Clinical and genetic heterogeneity. (1984) *J. Med. Genet.* **21 (6):** 421–428.

Cotton, R.G. Current methods of mutation detection. (1993) *Mutat. Res.* **285 (1):** 125–144.

Davies, K.E., Read, A.P. (1992) Genes and markers. In Rickwood, D., Male, D. (eds) *Molecular Basis of Inherited Disease*. pp. 1–14. Oxford University Press, Oxford.

Dryja, T.P., Mcgee, T.L., Reichel, E., Hahn, L.B., Cowley, G.S., Yandell, D.W., Sandberg, M.A., Berson, E.L. A point mutation of the rhodopsin gene in one form of retinitis pigmentosa. (1990) *Nature* **343**: 364–366.

Evans, K., Al-Maghtheh, M., Fitzke, F.W., Moore, A.T., Jay, M., Inglehearn, C.F., Arden, G.B., Bird, A.C. Bimodal expressivity in dominant retinitis pigmentosa genetically linked to chromosome 19q. (1995) *Br. J. Ophthalmol.* **79 (9):** 841–846.

Farrar, G.J., Jordan, S.A., Kenna, P., Humphries, M.M., Kumarsingh, R., McWilliam, P., Allamand, V., Sharp, E., Humphries, P. Autosomal dominant retinitis-pigmentosa – localization of a disease gene (RP6) to the short arm of chromosome-6. (1991) *Genomics.* **11**: 870–874.

Farrar, G.J., Jordan, S.A., Kumar-Singh, R., Inglehearn, C.F., Gal, A., Gregory, C., Al-Maghtheh, M., Kenna, P.F., Humphries, M.M., Sharp, E.M., Shiels, P.M., Bunge, S., Hargrave, P.A., Denton, M.J., Schwinger, E., Bhattacharya, S.S., Humphries, P. (1994) Extensive genetic heterogeneity in autosomal dominant retinitis pigmentosa. In Hollyfield, J.G., Anderson, R.E., LaVail, M.M. (eds) *Retinal Degeneration. Clinical and Laboratory Applications.* pp. 63–77, Plenum Publishing Corporation, New York.

Greenberg, J., Goliath, R., Beighton, P., Ramesar, R. A new locus for autosomal dominant retinitis pigmentosa on the short arm of chromosome 17. (1994) *Hum. Mol. Genet.* **3**: 915–918.

Grondahl, J. Estimation of prognosis and prevalence of retinitis pigmentosa and Usher syndrome in Norway. (1987) *Clin. Genet.* **31 (4):** 255–264.

Humphries, P., Kenna, P., Farrar, G.J. (1994) Autosomal dominant retinitis pigmentosa. In Wright, A.F., Jay, B. (eds) *Molecular Genetics of Inherited Eye Disorders*. pp. 153–172, Harwood Academic Publishers, Chur, Switzerland.

Inglehearn, C.F., Bashir, R., Lester, D.H., Jay, M., Bird, A.C., Bhattacharya, S.S. A 3-bp deletion in the rhodopsin gene in a family with autosomal dominant retinitis pigmentosa. (1991) *Am. J. Hum. Genet.* **48**: 26–30.

Inglehearn, C.F., Keen, T.J., Bashir, R., Jay, M., Fitzke, F., Bird, A.C., Crombie, A., Bhattacharya, S. A completed screen for mutations of the rhodopsin gene in a panel of patients with autosomal dominant retinitis pigmentosa. (1992) *Hum. Mol. Genet.* **1**: 41–45.

Inglehearn, C.F., Carter, S.A., Keen, T.J., Lindsey, J., Stephenson, A.M., Bashir, R., Al-Maghtheh, M., Moore, A.T., Jay, M., Bird, A.C. A new locus for autosomal dominant retinitis pigmentosa on chromosome 7p. (1993) *Nat. Genet.* **4:** 51–53.

Jeffreys, A.J., Wilson, V., Thein, S.L. Hypervariable 'minisatellite' regions in human DNA. (1985) *Nature* **314 (6006):** 67–73.

Jordan, S.A., Farrar, G.J., Kenna, P., Humphries, M.M., Sheils, D.M., Kumar-Singh, R., Sharp, E.M., Soriano, N., Ayuso, C., Benitez, J. Localization of an autosomal dominant retinitis pigmentosa gene to chromosome 7q. (1993) *Nat. Genet.* **4:** 54–58.

Keen, T.J., Inglehearn, C.F., Kim, R., Bird, A.C., Bhattacharya, S.S. Retinal pattern dystrophy associated with a 4 bp insertion at codon 140 in the RDS-peripherin gene. (1994) *Hum. Mol. Genet.* **3:** 367–368.

Keen, T.J., Inglehearn, C.F., Green, E.D., Cunningham, A.F., Patel, R.J., Peacock, R.E., Gerken, S., White, R., Wessenbach, J., Bhattacharya, S.S. YAC contig spanning the dominant retinitis pigmentosa locus (RP9) on chromosome 7p. (1995) *Genomics* **28:** 383–388.

Kumar-Singh, R., Jordan, S.A., Farrar, G.J., Humphries, P. Poly (T/A) polymorphism at the human retinal degeneration slow (RDS) locus. (1991) *NAR* **19:** 5800.

Lathrop, G.M., Lalouel, J.M., Julier, C., Ott, J. Strategies for multilocus linkage analysis in humans. (1984) *Proc. Nat. Acad. Sci. USA* **81 (11):** 3443–3446.

McWilliam, P., Farrar, G.J., Kenna, P., Bradley, D.G., Humphries, M.M., Sharp, E.M., McConnell, D.J., Lawler, M., Sheils, D., Ryan, C. Autosomal dominant retinitis pigmentosa (ADRP): localization of an ADRP gene to the long arm of chromosome 3. (1989) *Genomics* **5:** 619–622.

Nichols, B.E., Sheffield, V.C., Vandenburgh, K., Drack, A.V., Kimura, A.E., Stone, E.M. Butterfly-shaped pigment dystrophy of the fovea caused by a point mutation in codon 167 of the RDS gene. (1993) *Nat. Genet.* **3:** 202–207.

Orita, M., Suzuki, Y., Sekiya, T., Hayashi, K. Rapid and sensitive detection of point mutations and DNA polymorphisms using the polymerase chain reaction. (1989) *Genomics* **5:** 874–879.

Ott, J. (1986) A short guide to linkage analysis. In Davies, K.E. (ed.) *Human Genetic Diseases a Practical Approach.* pp. 19–32. IRL Press, Oxford and Washington, DC.

Parimoo, S., Kolluri, R., Weissman, S.M. cDNA selection from total yeast DNA containing YACs. (1993) *NAR.* **21 (18):** 4422–4423.

Weber, J.L., May, P.E. Abundant class of human DNA polymorphisms which can be typed using the polymerase chain reaction. (1989) *Am. J. Hum. Genet.* **44:** 388–396.

Wells, J., Wroblewski, J., Keen, J., Inglehearn, C., Jubb, C., Eckstein, A., Jay, M., Arden, G., Bhattacharya, S., Fitzke, F. Mutations in the human retinal degeneration slow (RDS) gene can cause either retinitis pigmentosa or macular dystrophy. (1993) *Nat. Genet.* **3:** 213–218.

Xu, S.Y., Schwartz, M., Rosenberg, T., Gal, A. A ninth locus (RP18) for autosomal dominant retinitis pigmentosa maps in the pericentromeric region of chromosome 1. (1996) *Hum. Mol. Genet.* **5:** 1193–1197.

Chapter 20
Problems in dealing with linkage heterogeneity in autosomal recessive forms of retinitis pigmentosa

Alan F. Wright, Peter W. Teague,
Elspeth Bruford and Andrew Carothers

Introduction

The degenerative diseases of the retina collectively termed retinitis pigmentosa (RP) are a group of conditions characterised by pigmentary retinopathy, reduced or absent electroretinogram, night blindness and progressive constriction of the visual fields, commonly leading to severe visual impairment by middle age (Heckenlively, 1988). The prevalence of RP in the general population is approximately 1 in 3,000. Ten years ago, it was appreciated that there were autosomal dominant, recessive and X-linked forms of RP, but only within the last few years has the true extent of the genetic heterogeneity underlying this condition become apparent. Within the common group of 'non-syndromal' forms of RP, in which the retina is the only organ affected, 18 loci have been genetically or physically mapped and many more remain to be found (Dryja and Li, 1995). Within the 'syndromal' forms of RP, in which the disorder affects both the retina and other organs, as many as 100 different conditions are known, although most of them are rare (Sullivan and Daiger, 1996). However, the diversity within this group is indicated by the findings in two of the commoner syndromes, Usher's syndrome and Bardet–Biedl syndome, between which there is evidence for at least 12 different loci. This surprising extent of genetic heterogeneity has major implications for

GENETIC MAPPING OF DISEASE GENES
ISBN 0-12-232735-7

genetic counselling, treatment strategies and gene mapping, the last of which is the subject of this paper.

Within autosomal dominant RP, genetic mapping efforts have been highly successful. The general strategy is to identify a large kindred with sufficient informative meioses to provide good statistical support either for including or excluding a chromosomal region by linkage analysis. The population and phenocopy rates of RP are relatively low and diagnoses are usually clearcut so that standard linkage techniques serve very well. The penetrance is reduced in some kindreds, but if the family is large enough and enough markers can be typed, a positive result can be obtained, and the analysis is relatively straightforward. Few false-positive linkages have been reported in RP studies. It is, however, worth emphasising that of the eight dominant RP loci mapped to date, none of them have been mapped using a 'family pooling' linkage strategy. Access to large extended families is more problematic in autosomal recessive disorders, where pooling of data from small nuclear families has been widely used in less heterogeneous conditions. How are investigators coping with the genetic diversity in recessive forms of RP? Autosomal recessive RP loci have been genetically mapped using three approaches.

- Linkage analysis using pooled nuclear families from the general (outbred) population.
- Linkage analysis using pooled nuclear families from specific founder populations.
- Linkage analysis in single extended consanguineous kindreds.

Examples of each will be discussed and their relative merits contrasted in different situations.

Mapping genes in recessive RP

Non-syndromal RP

Patients showing autosomal recessive inheritance constitute one of the commoner subgroups of RP and yet only two recessive forms of non-syndromal RP have been mapped by linkage analysis (Knowles *et al.*, 1994; van Soest *et al.*, 1994), both using single extended inbred pedigrees (*Table 20.1*). Four other genes have been directly identified using a candidate gene/mutation analysis approach (Rosenfeld *et al.*, 1992; McLaughlin *et al.*, 1993; Dryja *et al.*, 1995; Huang *et al.*, 1995) (see *Table 20.1*). All of the specific genes have been identified by screening large numbers (>100) of independent patient samples for mutation using

Table 20.1. Summary of genetic loci mapped or identified in patients with autosomal recessive RP, including some of the commoner syndromal forms

Genetic type	Locus	Chromosomal location	Identified gene	References
Autosomal recessive	RHO	3q21-q24	Rhodopsin	Rosenfeld et al., 1992
	PDEB	4p16	Rod cyclic GMP-phosphodiesterase β subunit	McLaughlin et al., 1993
	PDEA	5q31.2-q34	Rod cyclic GMP phosphodiesterase α subunit	Huang et al., 1995
	CNCG	4p14-q13	Rod cyclic GMP-gated channel α subunit	Dryja TP et al., 1995
	RP12 (PPRPE)	1q31-q32.1	NK	van Soest et al., 1994 van den Born et al., 1994
	RP14	6p21.3	NK	Knowles et al., 1994 Shugart et al., 1995
Autosomal recessive syndromes	Usher's syndrome type 1 (USH1A)	14q32	NK	Kaplan et al., 1992
	(USH1B)	11q13.5	Myosin VIIA	Kimberling et al., 1992 Weil et al., 1995
	(USH1C)	11p15.2-p14	NK	Smith et al., 1992
	Usher's syndrome type 2 (USH2A)	1q42-qter	NK	Kimberling et al., 1990
	Bardet–Biedl syndrome (BBS1)	11q13	NK	Leppert et al., 1994
	(BBS2)	16q21	NK	Kwitek-Black et al., 1993
	(BBS3)	3p13-p12	NK	Sheffield et al., 1994
	(BBS4)	15q22.3-q23	NK	Carmi et al., 1995
	Leber's congenital amaurosis	17p	NK	Camuzat et al., 1995
	Refsum's disease	10p	Phytanic acid oxidase	Nadal et al., 1995
	Gyrate atrophy	10q26	Ornithine-δ-aminotransferase	Brody et al., 1992
	Oguchi's disease	2q	S-arrestin	Fuchs et al., 1995
	Abetalipo-proteinaemia	4q22-q24	Microsomal triglyceride transfer protein	Narcisi et al., 1995

candidate genes concerned with the visual transduction pathway. The proportion of recessive patients accounted for by these genes is, however, quite low. The α and β subunits of the rod photoreceptor enzyme cyclic GMP phosphodiesterase and rod cyclic GMP-gated channel α subunit account for about 1.2%, 6% and 1.7% of recessive patients respectively (Dryja and Li, 1995; Huang *et al.*, 1995). Rhodopsin mutations also account for less than 1% of recessive RP patients in most populations, compared with 25–30% of dominant RP patients (Rosenfeld *et al.*, 1992).

The two recessive non-syndromal RP loci identified by linkage mapping have relied on mapping within single extended pedigrees (Knowles *et al.*, 1994; van Soest *et al.*, 1994). Large inbred kindreds with recessive RP are uncommon, but the general experience in using them has been favourable. The pitfalls arising from pooling of families in the presence of genetic heterogeneity are generally avoided. This is not always the case, however; a small inbred Dutch community in which there was a high prevalence of recessive RP was found on closer examination to contain at least two clinically distinct forms of recessive RP (van den Born *et al.*, 1994). One of these clinical subtypes (RP12) with para-arteriolar preservation of the retinal pigment epithelial (PPRPE) was mapped to chromosomal region 1q31-q32.1 by linkage analysis, but only when families with this distinct phenotype were recognised and analysed separately (van Soest *et al.*, 1994). The other clinical subtype remains to be mapped. The second recessive RP locus to be mapped by linkage, RP14, was mapped to chromosomal region 6p21.3 in a large consanguineous Dominican kindred (Knowles *et al.*, 1994; Shugart *et al.*, 1995).

Recessive RP in the Sardinian population The most extreme example of a founder effect producing a high frequency of RP within an isolate population is in the remote island of Tristan da Cunha, where the estimated carrier frequency of autosomal recessive RP was found to be as high as 1 in 5 (Roberts, 1980). Autosomal recessive RP is also unusually prevalent in the island of Sardinia, where it accounts for 77% of families in which the mode of inheritance can be established (Fossarello *et al.*, 1993), comparable only with a Swiss isolate population (Amman *et al.*, 1961). The overall prevalence of RP is not greatly increased in the island, being estimated to be 1 in 3355 in the south-central area (Fossarello *et al.*, 1993). However, the over-representation of autosomal recessive and under-representation of dominant and X-linked forms suggests that there may be less genetic heterogeneity within the population, which could facilitate the genetic mapping of this heterogeneous disorder.

The island of Sardinia contains a genetically isolate population of about 1.6 million people, which has, like Finland and other isolates, an unusual spectrum of genetically influenced disorders, probably resulting from genetic drift in a small founder population. It is one of the most extreme genetic outliers within the European population, together with the Lapp, Icelandic and Basque populations (Cavalli-Sforza *et al.*, 1994). The earliest inhabitants of Sardinia date back to 6,000–9,000 years ago when it was settled from Southern Europe, and when the population size is estimated to have been less than 2,000 individuals (Cavalli-Sforza *et al.*, 1994). Large genetic drift effects are likely to have occurred in this period up until the population expansion of the Nuragic era (1,500–400 B.C.) when the population size may have risen to about 200,000, freezing further major genetic drift effects. The genetic influence of further coastal settlements, particularly by the Phoenicians in the ninth century is still evident, but in general there has been little immigration into the island in the past 2,000 years (Cavalli-Sforza *et al.*, 1994). Consanguinity rates in Sardinia vary widely, but are particularly high in villages at higher altitudes where they reach rates up to 10% due to a combination of longstanding cultural and geographical factors (Workman *et al.*, 1975). The frequencies of insulin-dependent diabetes mellitus, multiple sclerosis, Wilson's disease and glaucoma are significantly higher than in most populations, while other diseases are less common (Bernardinelli *et al.*, 1994). The β-globin mutation (β^{39}) associated with 95% of β-thalassaemia chromosomes in Sardinia is rare elsewhere, and extreme frequencies of various human leucocyte antigen (HLA), blood group and enzyme markers are found. It is therefore a plausible assumption that autosomal recessive RP could show an unusual and less diverse spectrum of mutations than in more outbred populations.

The first attempts at genetic mapping recessive genes in the Sardinian population were focused on a set of 11 nuclear families, with 2–4 affected members per family (Fossarello *et al.*, 1993; Wright *et al.*, 1995). Simulations showed that a single gene accounting for the disease in 35% of these families would be detectable with a logarithm of odds (lod) score of 3.3, given tightly linked markers. One hundred and ninety five markers were typed, spanning most human chromosomes, but no consistent linkage was found in the total data set. The chromosomal regions screened for linkage encompassed known candidate genes, and markers were generally spaced at 20 centiMorgan (cM) intervals. Only one interesting linkage observation was made. A small subset of five families showed no recombination with a marker on chromosome 14, D14S80. Heterogeneity analysis suggested the possibility of linkage heterogeneity, but the result was of borderline significance. It was then noted that one of the linked families showed a region of homozygosity extending over three adjacent markers (D14S262, D14S64, D14S80) in all three affected offspring of a second-cousin marriage (Wright *et al.*, 1995). This raised the possibility that all three inherited an identical by descent (IBD) region surrounding the disease gene from a common ancestor. However, none of the other linked families showed homozygosity spanning this whole region, which covers about 6 cM. The multipoint lod score in the family was 0.9–2, depending on the gene frequency assumptions and whether it was calculated using homozygosity mapping or standard linkage programmes.

The result was felt to be sufficiently interesting to warrant the analysis of a candidate gene located within the homozygous region. This gene was the neural retina leucine zipper (NRL) gene, which is on the same cosmid clone as D14S64, binds to the rhodopsin promoter, and may regulate its expression (Rehemtulla *et al.*, 1996). Mutation screening using single strand conformational polymorphism (SSCP) analysis and direct sequencing of the entire coding region failed to identify any variants (Bruford *et al.*, in press). It remains possible that a mutation escaped detection, but the less exciting possibility remains that homozygosity of three adjacent microsatellite loci, in this case representing the most common haplotype in that population, had occurred by chance rather than as a result of IBD. Three children of a second cousin marriage share only 0.1% (3 cM) of their genomes that are homozygous by descent (HBD) and the average size of an HBD segment shared between three affected children around the disease locus is 16.7 cM (*see* Appendix). Therefore if a true HBD segment is confirmed within the family by identification and typing of new markers within the 6 cM segment there is a reasonable probability that the gene is located within the HBD segment.

Summary The most predictable approach to mapping autosomal recessive RP genes is to use extended consanguineous kindreds, which largely circumvents the problem of genetic heterogeneity. The use of linkage analysis in pooled nuclear families with recessive RP has been successful in the genetic mapping of two forms of syndromal RP (see below), but has not yet been successful in identifying genes in non-syndromal RP. This strategy is uncertain in the light of the presumed genetic heterogeneity, even in isolate populations such as Sardinia. It has been successfully used in the situation where clinical criteria can identify subgroups of families that are genetically homogeneous.

Syndromal RP

Syndromal forms of recessive RP are a diverse group and most of the conditions are individually rare. The commonest is Usher's syndrome, in which RP occurs together with deafness; it affects 6–10% of RP patients and 3–6% of the deaf population (Heckenlively, 1988).

Usher's syndrome The 'standard' method of linkage mapping using pooled nuclear families from a general population was used to map the locus (USH2A) for type II Usher's syndrome, which is characterised by a moderately severe congenital hearing impairment together with a later onset pigmentary retinopathy (Kimberling *et al.*, 1994). This subtype of Usher's syndrome, which affects about 50% of all patients with the syndrome, is genetically heterogenous, but most families map to the 1q42-qter region (Kimberling *et al.*, 1990). In type I Usher's syndrome, which is characterised by profound congenital hearing loss and vestibular deficits in addition to RP, there are mutations in at least three different loci (14q32, 11q13.5, 11p15.2-p14), one of which has been specifically identified as myosin VIIA (see *Table 20.1*).

The USH1A and USH1C loci were identified by linkage analysis carried out within specific subpopulations: the French Poitou–Charentes and North American Acadian populations, respectively (Kaplan *et al.*, 1992; Smith *et al.*, 1992). The eight Poitou–Charentes families showing linkage to 14q32 were not known to be related, but the geographic clustering of USH1A families, so far exclusively found in this region, suggests a founder effect. Usher's syndrome shows an increased prevalence in the Acadian population, suggesting that the French Acadians who were originally from Normandy and underwent forced migration from Quebec to the bayous of Louisiana about 200 years ago and remained as a genetic isolate until the last 50 years may have an increased population frequency of an USH1C mutation as a result of a founder effect (Smith *et al.*, 1992). An interesting point to note is that despite their restricted genetic origin, both types of Usher's syndrome are present in this population, with only type 1 showing an increased prevalence (Kimberling *et al.*, 1994).

Type 3 Usher's syndrome is a rare form in which there is non-congenital progressive hearing loss and has also been mapped using pooled family data by linkage analysis in the Finnish population, which is another genetic isolate (Sankila *et al.*, 1995).

The relative frequency of USH1A, B and C mutations in Usher's syndrome, which affects 3–4/100,000 in most populations is unknown, but it appears that the myosin VIIA gene (USH1B) is the most common cause (Weil *et al.*, 1995).

Bardet–Biedl syndrome The strategy of mapping a single extended kindred has proved very successful in Bardet–Biedl syndrome (BBS), an autosomal recessive disorder characterised by RP, polydactyly, hypogenitalism, obesity and mental retardation (Green *et al.*, 1989). This is a less common form of syndromal RP, affecting 1 in 13,500–1 in 160,000 in different populations. The comparative rarity and unusual combination of phenotypic features initially suggested that the condition might result from mutation at a single locus, but this has certainly not proved to be the case. Three BBS genes were mapped in relatively quick succession to chromosomal regions 16q21 (BBS2), 3p13-p12 (BBS3) and 15q22.3-q23 (BBS4) by an Iowa group using samples from three extended consanguineous Bedouin kindreds (Kwitek-Black *et al.*, 1993; Sheffield *et al.*, 1994; Carmi *et al.*, 1995). The BBS4 locus was identified by homozygosity mapping using an interesting pooled DNA sample strategy (Carmi *et al.*, 1995). Another BBS locus was identified in chromosomal region 11q13 (BBS1) within a subset of 17 of 31 North American BBS families (Leppert *et al.*, 1994). In order to investigate the relative frequency of these loci within a population of predominantly European BBS families, we undertook a linkage study using 37 polymorphic markers in 29 BBS families, each containing at least two affected members (Bruford *et al.*, 1997). The results showed clear evidence of linkage to the chromosome 11 (BBS1) and 15 (BBS4) loci, probable linkage to the chromosome 16 (BBS2) locus, no detectable linkage to the chromosome 3 locus (BBS3), and a further group showing no linkage to any of the known loci (*Tables 20.2, 20.3*).

Table 20.2. Results of linkage heterogeneity analysis in 29 Bardet–Biedl families, with each locus considered independently.

Locus	θ (s.e.)	lod score	α (%) (95% CI)	Linkage hypotheses	χ^2	df	p value
BBS1	66 (1.1)	6.26	56% (30–77)	H_2 vs H_0 H_2 vs H_1	28.84 6.64	2 1	0.0000 0.0050
BBS4	94 (1.3)	6.09	35 (14–57)	H_2 vs H_0 H_2 vs H_1	28.07 10.39	2 1	0.0000 0.0006
BBS2	76 (3.6)	1.09	27 (3–97)	H_2 vs H_0 H_2 vs H_1	5.01 3.60	2 1	0.0453 0.0290

The genetic locations (θ) of the BBS1 (11q13), BBS4 (15q22.3-q23) and BBS2 (16q21) loci are shown in centiMorgans (cM) from a fixed reference point (distal short arm of the chromosome), with standard errors (s.e.). Alpha (α) is the proportion of families linked to each locus which is given together with its 95% confidence interval (CI). The hypotheses considered are H_2 (one linked, one unlinked locus), H_1 (all families linked to one locus) and H_0 (all families unlinked) (df, degrees of freedom)

Table 20.3. Estimates of the proportion of Bardet–Biedl families linked (α) to one, two or three loci, corresponding to BBS1 (11q13), BBS4 (15q22.3-q23) and BBS2 (16q21).

	θ_1 (cM)	α_1 (%)	θ_2 (cM)	α_2 (%)	θ_3 (cM)	α_3 (%)	Log$_e$ likelihood	Linkage hypotheses	χ^2	df	p value
H_2	66	56					14.42	H_2 vs H_0	28.84	2	0.0000
H_4	66	40	94	32			23.69	H_4 vs H_2	18.54	2	0.0000
H_5	66	36	94	32	76	24	26.04	H_5 vs H_4	4.70	2	0.0477

The genetic locations of each locus are given in centiMorgans (cM) from a fixed reference point and the log$_e$ likelihoods and statistical support for each hypothesis, as shown
H_0 No linkage; $\theta = 0.5$, $\alpha = 1$
H_1 All families linked ; $\theta < 0.5$, $\alpha = 1$
H_2 Some families linked; $\theta < 0.5$, $\alpha < 1$
H_4 Two family groups linked; $\theta_1 < 0.5$, $\theta_2 < 0.5$, $\alpha_1 < 1$, $\alpha_2 < 1$
H_5 Three family groups linked; $\theta_1 < 0.5$, $\theta_2 < 0.5$, $\theta_3 < 0.5$, $\alpha_1 < 1$, $\alpha_2 < 1$, $\alpha_3 < 1$

The statistical analysis was complex in view of the extent of heterogeneity. Multipoint LOD scores were derived for markers spanning each known BBS locus using the LINKAGE (FASTLINK) program (Cottingham *et al.*, 1993) by means of serial three-point analyses on adjacent loci, moving along by one locus at a time. A linear genetic map over equally spaced points was derived from these LOD scores and used for linkage heterogeneity analysis either using the HOMOG program package (Terwilliger and Ott, 1994) or an equivalent Fortran program (Teague *et al.*, 1994). Multilocus heterogeneity analysis was carried out, initially testing for linkage to a single locus in the presence of an unlinked group, for each set of chromosomal markers in turn. The results are shown in *Table 20.2*. An estimated 56% of families were linked to BBS1 in 11q13 with a maximum LOD score of 6.26 under the hypothesis of linkage heterogenity. Similarly, 35% of families showed linkage to the BBS4 locus in 15q22.3-q23 with a maximum LOD score of 6.1. No detectable linkage to BBS3 was found, but there was weak evidence of linkage in 27% of families to BBS2 in 16q21 at a marginal level of significance (see *Table 20.2*).

It is clear from *Table 20.2* that adding the estimated percentages of linked families (α) for each chromosomal locus taken separately results in a total exceeding 100%. However, the estimates of α were much less precise than those of the genetic locations (Θ). To obtain an unbiased estimate of how the disease loci were distributed between families it was necesssary to consider all the loci at once. This was achieved by combining the three multipoint genetic maps into one: Ch11|Ch15|Ch16. Hypothesis tests were then carried out on the 'pseudochromosome' appropriate to one, two and three linked loci (H_2, H_4, H_5). The results are shown in *Table 20.3*, where H_2 finds a single locus (BBS1) on chromosome 11 ($\chi^2 = 28.84$, 2 df). Allowing for two linked loci (H_4) gives a better fit to the data and adds a chromosome 15 (BBS4) locus ($\chi^2 = 18.54$, 2 df). Allowing for three linked loci (H_5) adds the chromosome 16 locus (BBS2), but this time at a marginal level of significance ($\chi^2 = 4.70$, 2 df). The positions of the three loci agree well with the single chromosome analysis. Estimates for the percentage of families linked to each locus (α) agree well with the previous estimates for the chromosome 15 and 16 loci: for the chromosome 11 locus (36% versus 56%) the new estimate lies within the previous confidence interval (see *Table 20.3*).

The reason for the lower statistical confidence of the BBS2 linkage, even although it represents a similar proportion of BBS families to the BBS4 group, appears to be the higher proportion of small or non-consanguineous families in the former group. Finally, in an estimated 8% of families, all four loci could be excluded. The genetic locations for the BBS1and BBS2 loci were consistent with but did not refine those published previously. An improvement in the genetic mapping of BBS4 was possible, however, within one of the consanguineous families. A total of six of 29 BBS families showed parental consanguinity and these proved to be by far the most useful, in view of the increased statistical certainty of linkage where homozygosity of adjacent markers was observed. This results from indirect tracking of the disease gene through many additional meioses from the common ancestor. For example, three consanguineous families, each with either two or three affected children, showed homozygosity for three adjacent markers (D15S125, D15S204, D15S131) consistent with IBD and resulting in significantly higher LOD scores (1.69, 2.37, 2.85) than in similar sized non-consanguineous families. Two crossovers in one of these families narrowed the BBS4 interval from 7.5 cM (Carmi *et al.*, 1995) to about 2 cM, which is within the scope of a positional cloning effort. Since the maximum lod score in this family was 2.85, the recombinant haplotypes provide a reliable basis for map refinement.

Summary The unexpectedly high level of genetic heterogeneity in BBS creates problems in the analysis of a mixed family sample: at least four loci were detected in a sample of 29 predominantly European BBS families. Two loci were detected with a high level of statistical certainty, in 36% (BBS1) and 32% (BBS4) of families respectively, while the BBS2 (24%) and unlinked (8%) groups were less robust findings. A much larger sample of families or a group in which consanguinity rates are higher may be required, to convincingly detect linkage in a subset representing less than about one-third of families. The most useful group of families were those with parental consanguinity, so even in the absence of extended kindreds, consanguineous families with only two or three affected members can provide strong evidence for linkage to a particular locus and useful fine mapping data (a point not emphasised in the literature).

RP as a model for mapping genetically complex disorders

One of the most consistent findings in RP research has been the observation that about one-half of all patients have no family history (the simplex group). Clearly many of these may be autosomal recessive, but have no affected relatives as a result of small family size and the expected segregation ratio of 0.25. However, when segregation analyses have been carried out in RP populations, there is a clear excess of simplex patients; as many as 20% of patients are sporadic and not accounted for within the major Mendelian categories (Boughman and Caldwell, 1982; Jay, 1982). Some of this group may have new dominant mutations, but this proportion is expected to be low since the reproductive fitness is high (Bundey and Crews, 1984). Undetected X-linkage, dominant genes with reduced penetrance, and probably a small number of phenocopies may also contribute. However, digenic inheritance of RP has been reported (Kajiwara *et al.*, 1994) and the possibility that a sizeable group of RP patients may show an oligogenic mode of inheritance should perhaps be seriously considered. On this view, RP is a diverse series of retinal degenerations in which aetiological factors can be represented on a genetic-environmental continuum. Fully penetrant Mendelian disorders account for the disease at one extreme and phenocopies at the other, with a significant oligogenic group in between. Interactions between susceptibility genes and environmental factors such as light exposure or vitamin intake may also be important. If this is the case, attempting to map the 'weak' or interaction-dependent genes might be seen as unnecessarily difficult and uncertain. Although this may be so, the situation is in some respects more favourable than in the detection of genes contributing to genetically complex disorders such as diabetes mellitus, asthma or the major psychoses, which are now the fashionable objects of study. The favourable features are as follows:

- Diagnosis is straightforward and clearcut.
- Onset age is relatively early and homogeneous.
- The frequency of simple phenocopies is probably low.
- The known diversity of susceptibility genes, many of them interacting within a common (phototransduction) pathway.
- Examples of digenic inheritance and genes with reduced penetrance are established features of the disease, making an oligogenic hypothesis relatively well supported.
- The population prevalence of RP is at least an order of magnitude lower than the common oligogenic disorders, which facilitates genetic mapping.

How can this group of 'oligogenic RP' patients be studied and the susceptibility genes mapped if they are best identified by a lack of family history? First, allelic association studies can be carried out. This method is not strongly affected by misspecifying the mode of inheritance. The ideal population is a founder one such as the one in Sardinia, since the genetic diversity is likely to be reduced. This is best carried out by typing affected individuals for a series of candidate genes, establishing haplotypes in affected individuals from the parental genotypes, and comparing these with the non-transmitted 'control' haplotypes, the so-called haplotype relative risk method (Falk and Rubinstein, 1987). Provided that there are good candidate genes, it is reasonable to ask whether these loci contribute to susceptibility within a sample of simplex patients. This would be a feasible study using a variety of phototransduction or other candidate eye genes, whereas a thorough mutation analysis of the same genes, including promoter and other regulatory regions (which may be over-represented in 'weak' susceptibility loci), without first acquiring genetic evidence for a causal role, would be prohibitive. To extend such a study design to detecting unknown susceptibility genes by linkage is a much less attractive prospect. The disease mutations within the founder population may be as old as 2,000 years or more so surrounding IBD segments will be too small for efficient mapping by searching for regions of homozygosity.

Discussion

The best method of mapping recessive disease genes in the presence of extensive genetic heterogeneity is undoubtedly to use linkage analysis in extended consanguineous kindreds. These have already provided reliable information on the mapping of genes both in non-syndromal and syndromal forms of RP. However, such families are not readily available and may not even be representative of the disorder in less restricted populations. Extended kindreds generally have sufficient meioses to map the disease gene to within a 10–20 cM region, but insufficient to fine map it to the point where a positional cloning project can be contemplated. Bridging this gap, from 10 cM to 1 cM, is difficult with recessive disease in the presence of heterogeneity. One alternative is to study a founder population, either using genetic linkage or a genetic association study.

A genetic linkage study of 11 nuclear families from Sardinia with autosomal recessive RP was carried out using 195 genetic markers without identifying any certain linkages. This result led us to think that even with the restricted genetic heterogeneity expected within the Sardinian population, linkage only reliably gives a positive mapping within single extended consanguineous kindreds. A further possibility would be to use genetic markers to search for common regions of IBD, as indicated by regions of homozygosity covering adjacent markers, within apparently unrelated affected individuals. However, the historical and archaeological evidence suggests that founder disease genes may have originated as far back as 2,000 years ago, when population expansion probably froze further large-scale genetic drift effects. Assuming 25 years per generation, the average length of a chromosomal segment persisting around a disease gene for this duration is only 1–2 cM, which is too small to map efficiently using genetic markers. This may be the optimum strategy for mapping mutations arising within the last 150–350 years (te Meerman et al., 1995), but this is unlikely to be the case within Sardinia where the proportion of autosomal recessive RP appears to be higher than in outbred populations throughout the island and not merely confined to small geographic regions. The situation in Tristan da Cunha may fit into this category very precisely, since a founder settling on the island within the last 200 years was clearly a carrier for recessive RP and the present population derives from only 15 founders (Roberts, 1980).

Isolate populations such as Sardinia are, however, suitable for analysing candidate genes for homozygosity using close flanking or intragenic markers. A high proportion of genes concerned with the visual transduction process has been cloned so there are more excellent candidates for retinal degenerations than virtually any other physiological system in man. Mutations in many of these genes have already been shown to cause retinal degenerations either in man, transgenic or naturally occurring mouse mutants (reviewed by Dryja and Li, 1995). A strategy to look for genetic association between the candidate gene alleles transmitted from parents to affected individuals compared with the non-transmitted parental alleles is both technically feasible and scientifically justifiable. A recent study by Genin and Clerget-Darpoux (1996) showed that while association studies in random mating populations are well supported theoretically, the situation in inbred populations is also favourable for such studies and in many situations inbreeding greatly facilitates the detection of disease associations. The same type of gene association study in a more diverse outbred population would be in danger of encountering a prohibitive level of genetic heterogeneity.

Once disease loci have been identified, it is possible to establish their relative contributions to disease within a series of nuclear families from outbred or ethnically diverse populations. Three BBS loci were detected in a series of 29 predominantly nuclear families, although the consanguineous families, even with only two or three affected members, were much the most informative in terms of mapping information.

In summary, the lessons to be learnt from our experience of mapping a genetically diverse disorder such as autosomal recessive RP are as follows.

- Extended inbred kindreds are best for the initial mapping.
- Once the major loci have been detected, the relative frequencies of the different genes can be confirmed in a series of about 30 nuclear families, provided the locus accounts for disease in at least 30% of families, although consanguineous families provide the most information.
- In the absence of extended pedigrees, isolate populations are well suited for screening of candidate genes either by genetic linkage or, if families only have single affected members, by genetic association.
- Genetic association studies using evenly spaced markers may be useful for primary identification of a recessive disease gene in the special, and perhaps

unusual, situation where there is good evidence for a founder gene being introduced into an isolate population within the past 150–350 years.

Acknowledgements

We would like to acknowledge the British Retinitis Pigmentosa Society and the Gift of Thomas Pocklington for financial support in these studies.

Appendix

Regions of homozygosity by descent

Suppose an individual is homozygous by descent (HBD) at a particular locus, A say (e.g. a recessive disease), and that the region of HBD extends a distance x cM in one direction from A to locus B. Let n_m, n_p denote respectively the numbers of meioses from the founder in the maternal and paternal lines. Now suppose the distance AB is divided into a large number, n say, of small segments each of length x/n cM. Then

$Pr(x).\delta x = Pr$(no recombination in any of the individual segments)

$X Pr$ (at least one recombination between x and $x + \delta x$ in either the maternal or paternal line)

$$\left[1 - \frac{x}{100n}\right]^{n(n_m + n_p)} \times \left[1 - \left[1 - \frac{\delta x}{100}\right]^{n_m + n_p}\right] \qquad \textbf{(Eq. 20.1)}$$

$$\rightarrow e^{\frac{-(n_m + n_p)x}{100}} \cdot \frac{n_m + n_p}{100} \cdot \delta x$$

as $n \rightarrow \infty$ and $\delta x \rightarrow 0$

Hence x has an exponential distribution with probability distribution function (p.d.f.) $\alpha e^{-\alpha x}$ where:

$$\alpha = \frac{n_m + n_p}{100} \qquad \textbf{(Eq. 20.2)}$$

The total region of HBD, y say, is obtained by adding together the regions on either side of A. The sum of two independent exponential variates has (from standard statistical theory) a γ distribution with p.d.f. $\alpha y e^{-\alpha y}$. This has a mean value given by:

$$\mu = \frac{2}{\alpha} = \frac{200}{n_m + n_p} \qquad \textbf{(Eq. 20.3)}$$

For example, in the case of the progeny of a first-cousin mating, we have $n_m = n_p = 3$, and therefore $\mu = 33.3$ cM.

The result has a small positive bias since it ignores the possibility that A may be situated close to a telomere.

For a sibship with k affected sibs $(n_m + n_p)$ is replaced by $(n_m + n_p + 2k - 2)$. The corresponding value for s independent sibships is:

$$\sum_{i=1}^{s} (n_{mi} + n_{pi} + 2k_i - 2) \qquad \textbf{(Eq. 20.4)}$$

References

Amman, F., Klein, D., Bohringer, H.R. Resultats preliminaires d'une enquete sur la frequence et la distribution geographique des degenerescences tapeto-retiniennes en Suisse (étude de cinq cantons). (1961) *J. Genet. Hum.* **10**: 99–127.

Bernardinelli, L., Maida, A., Marinoni, A., Clayton, D., Romano, G., Montomoli, C., *et al. Atlas of Cancer Mortality in Sardinia 1983–1987.* pp. 1–117. (1994) FATMA, Rome.

Boughman, J.A., Caldwell, R.J. Genetic and clinical characterization of a survey population with retinitis pigmentosa. In Deantl, D.L. (ed.) *Clinical, Structural and Biochemical Advances in Hereditary Eye Disorders.* pp. 147–166, (1982) Alan R. Liss, New York.

Brody, L.C., Mitchell, G.A., Obie, C., Michaud, J., Steel, G., Fontaine, G., Robert, M.F., Sipila, I., Kaiser Kupfer, M., Valle, D. Ornithine delta-aminotransferase mutations in gyrate atrophy. Allelic heterogeneity and functional consequences. (1992) *J. Biol. Chem.* **267 (5)**: 3302–3307.

Bruford, E.A., Riit, R., Teague, P.W., Porter, K., Thomson, K.L., Moore, A.T., Jay, M., Warburg, M., Schinzel, A., Tommerup, N., Tornqvist, K., Rosenberg, T., Patton, J., Mansfield, D.C., Wright, A.F. Linkage mapping in 29 Bardet–Biedl syndrome families confirms loci in chromosomal regions 11q13, 15q22.3–q23 and 16q21. (1997) *Genomics* **40**. (In press.)

Bundey, S., Crews, S.J. A study of retinitis pigmentosa in the City of Birmingham. II Clinical and genetic heterogeneity. (1984) *J. Med. Genet.* **21 (6)**: 421–428.

Camuzat, A., Dollfus, H., Rozet, J.M., Gerber, S., Bonneau, D., Bonnemaison, M., Briard, M.L., Dufier, J.L., Ghazi, I., Leowski, C., *et al.* A gene for Leber's congenital amaurosis maps to chromosome 17p. (1995) *Hum. Mol. Genet.* **4 (8)**: 1447–1452.

Carmi, R., Rokhlina, T., Kwitek-Black, A.E., Elbedour, K., Nishimura, D., Stone, E.M., Sheffield, V.C. Use of a DNA pooling strategy to identify a human obesity syndrome locus on chromosome 15. (1995) *Hum. Mol. Genet.* **4 (1)**: 9–13.

Cavalli-Sforza, L.L., Menozzi, P., Piazza, A. *The History and Geography of Human Genes.* pp. 1–535. (1994) Princeton University Press, New Jersey.

Cottingham, R.W. Jr, Idury, R.M., Schaffer, A.A. Faster sequential genetic linkage computations. (1993) *Am. J. Hum. Genet.* **53** (1): 252–263.

Dryja, T.P., Finn, J.T., Peng, Y.W., Mcgee, T.L., Berson, E.L., Yau, K.W. Mutations in the gene encoding the alpha subunit of the rod cGMP-gated channel in autosomal recessive retinitis pigmentosa. (1995) *Proc. Nat. Acad. Sci. USA* **92** (22): 10177–10181.

Dryja, T.P. and Li, T. Molecular genetics of retinitis pigmentosa. (1995) *Hum. Mol. Genet.* **4 Spec No:** 1739–1743.

Falk, C.T., Rubinstein, P. Haplotype relative risks: an easy reliable way to construct a proper control sample for risk calculations. (1987) *Ann. Hum. Genet.* **51** (3): 227–233.

Fossarello, M., Serra, A., Mansfield, D., Wright, A., Loudianos, J., Pirastu, M., Orzalesi, N. Genetic and epidemiological study of autosomal dominant (ADRP) and autosomal recessive (ARRP) retinitis pigmentosa in Sardinia. In Hollyfield, J.G., Anderson, R.E. (eds) *Retinal Degeneration.* pp. 79–90. (1993) Plenum Press, New York.

Fuchs, S., Nakazawa, M., Maw, M., Tamai, M., Oguchi, Y., Gal, A. A homozygous 1-base pair deletion in the arrestin gene is a frequent cause of Oguchi disease in Japanese. (1995) *Nat. Genet.* **10** (3): 360–362.

Genin, E., Clerget-Darpoux, F. Association studies in consanguineous populations. (1996) *Am. J. Hum. Genet.* **58** (4): 861–866.

Green, J.S., Parfrey, P.S., Harnett, J.D., Farid, N.R., Cramer, B.C., Johnson, G., Heath, O., McManamon, P.J., O'Leary, E., Pryse Phillips, W. The cardinal manifestations of Bardet–Biedl syndrome, a form of Laurence–Moon–Biedl syndrome. (1989) *N. Engl. J. Med.* **321** (15): 1002–1009.

Heckenlively, J.R. *Retinitis Pigmentosa.* pp. 1–269. (1988) Lippincott Company, Philadelphia.

Holt, I.J., Harding, A.E., Petty, R.K., Morgan Hughes, J.A. A new mitochondrial disease associated with mitochondrial DNA heteroplasmy. (1990) *Am. J. Hum. Genet.* **46** (3): 428–433.

Huang, S.H., Pittler, S.J., Huang, X., Oliveira, L., Berson, E.L., Dryja, T.P. Autosomal recessive retinitis pigmentosa caused by mutations in the alpha subunit of rod cGMP phosphodiesterase. (1995) *Nat. Genet.* **11** (4): 468–471.

Jay, M. On the heredity of retinitis pigmentosa. (1982) *Br. J. Ophthalmol.* **66** (7): 405–416.

Kajiwara, K., Berson, E.L., Dryja, T.P. Digenic retinitis pigmentosa due to mutations at the unlinked peripherin/RDS and ROM1 loci. (1994) *Science* **264**: 1604–1608.

Kaplan, J., Gerber, S., Bonneau, D., Rozet, J.M., Delrieu, O., Briard, M.L., Dollfus, H., Ghazi, I., Dufier, J.L., Frezal, J. A gene for Usher syndrome type I (USH1A) maps to chromosome 14q. (1992) *Genomics* **14**: 979–987.

Kimberling, W.J., Weston, M.D., Moller, C., Davenport, S.L., Shugart, Y.Y., Priluck, I.A., Martini, A., Milani, M., Smith, R.J. Localization of Usher syndrome type II to chromosome 1q. (1990) *Genomics* **7** (2): 245–249.

Kimberling, W.J., Moller, C.G., Davenport, S., Priluck, I.A., Beighton, P.H., Greenberg, J., Reardon, W., Weston, M.D., Kenyon, J.B., Grunkemeyer, J.A. Linkage of Usher syndrome type I gene (USH1B) to the long arm of chromosome 11. (1992) *Genomics* **14**: 988–994.

Kimberling, W.J., Weston, M., Moller, C. Clinical and genetic heterogeneity of Usher syndrome. In Wright, A.F., Jay, B. (eds) *Molecular Genetics of Inherited Eye Disorders.* pp. 359–381. (1994) Harwood Academic, Switzerland.

Knowles, J.A., Shugart, Y., Banerjee, P., Gilliam, T.C., Lewis, C.A., Jacobson, S.G., Ott, J. Identification of a locus, distinct from RDS-peripherin, for autosomal recessive retinitis pigmentosa on chromosome 6p. (1994) *Hum. Mol. Genet.* **3**: 1401–1403.

Kwitek-Black, A.E., Carmi, R., Duyk, G.M., Buetow, K.H., Elbedour, K., Parvari, R., Yandava, C.N., Stone, E.M., Sheffield, V.C. Linkage of Bardet–Biedl syndrome to chromosome 16q and evidence for non-allelic genetic heterogeneity. (1993) *Nat. Genet.* **5**: 392–396.

Leppert, M., Baird, L., Anderson, K.L., Otterud, B., Lupski, J.R., Lewis, R.A. Bardet–Biedl syndrome is linked to DNA markers on chromosome 11q and is genetically heterogeneous. (1994) *Nat. Genet.* **7**: 108–112.

McLaughlin, M.E., Sandberg, M.A., Berson, E.L., Dryja, T.P. Recessive mutations in the gene encoding the beta-subunit of rod phosphodiesterase in patients with retinitis pigmentosa. (1993) *Nat. Genet.* **4**: 130–134.

Nadal, N., Rolland, M.O., Tranchant, C., Reutenauer, L., Gyapay, G., Warter, J.M., Mandel, J.L., Koenig, M. Localization of Refsum disease with increased pipecolic acidaemia to chromosome 10p by homozygosity mapping and carrier testing in a single nuclear family. (1995) *Hum. Mol. Genet*, **4 (10)**: 1963–1966.

Narcisi, T.M., Shoulders, C.C., Chester, S.A., Read, J., Brett, D.J., Harrison, G.B., Grantham, T.T., Fox, M.F., Povey, S., de Bruin, T.W., *et al.* Mutations of the microsomal triglyceride-transfer-protein gene in abetalipoproteinemia. (1995) *Am. J. Hum. Genet.* **57 (6)**: 1298–1310.

Rehemtulla, A., Warwar, R., Kumar, R., Ji, X., Zack, D.J., Swaroop, A. The basic motif-leucine zipper transcription factor Nrl can positively regulate rhodopsin gene expression. (1996) *Proc. Nat. Acad. Sci. USA* **93 (1)**: 191–195.

Roberts, D.F. Genetic structure and the pathology of an isolated population. in Eriksson, A.W. (ed.) *Population Structure and Genetic Disorders.* pp. 7–26. (1980) Academic Press, London.

Rosenfeld, P.J., Cowley, G.S., Mcgee, T.L., Sandberg, M.A., Berson, E.L., Dryja, T.P. A null mutation in the rhodopsin gene causes rod photoreceptor dysfunction and autosomal recessive retinitis pigmentosa. (1992) *Nat. Genet.* **1**: 209–213.

Sankila, E.M., Pakarinen, L., Kaariainen, H., Aittomaki, K., Karjalainen, S., Sistonen, P., de la Chapelle, A. Assignment of an Usher syndrome type III (USH3) gene to chromosome 3q. (1995) *Hum. Mol. Genet.* **4**: 93–98.

Sheffield, V.C., Carmi, R., Kwitek-Black, A., Rokhlina, T., Nishimura, D., Duyk, G.M., Elbedour, K., Sunden, S.L., Stone, E.M. Identification of a Bardet–Biedl syndrome locus on chromosome 3 and evaluation of an efficient approach to homozygosity mapping. (1994) *Hum. Mol. Genet.* **3**: 1331–1335.

Shugart, Y.Y., Banerjee, P., Knowles, J.A., Lewis, C.A., Jacobson, S.G., Matise, T.C., Penchas Zadeh, G., Gilliam, T.C., Ott, J. Fine genetic mapping of a gene for autosomal recessive retinitis pigmentosa on chromosome 6p21. (Letter). (1995) *Am. J. Hum. Genet.* **57**: 499–502.

Smith, R.J., Lee, E.C., Kimberling, W.J., Daiger, S.P., Pelias, M.Z., Keats, B.J.B., Jay, M., Bird, A.C., Reardon, W., Guest, M., Ayyagari, R., Hejtmancik, J.F. Localization of two genes for Usher syndrome type I to chromosome 11. (1992) *Genomics.* **14**: 995–1002.

Sullivan, L.S., Daiger, S.P. (1996) Inherited retinal degeneration: exceptional genetic and clinical heterogeneity. Molec. Med. Today. **2**: 380–386.

te Meerman, G., van der Meulen, M.A., Sandkuijl, L.A. Perspectives of identity by descent (IBD) mapping in founder populations. (1995) *Clin. Exp. Allergy* **25 (Suppl. 2)**: 97–102.

Teague, P.W., Aldred, M.A., Jay, M., Dempster, M., Harrison, C., Carothers, A.D., Hardwick, L.J., Evans, H.J., Strain, L., Brock, D.J.H., Bundey, S., Jay, B., Bird, A.C., Bhattacharya, S.S., Wright, A.F. Heterogeneity analysis in 40 X-linked retinitis pigmentosa families. (1994) *Am. J. Hum. Genet.* **55**: 105–111.

Terwilliger, J.D., Ott, J. *Handbook of Human Genetic Linkage*. (1994) Johns Hopkins University Press, Baltimore,

van den Born, L.I., van Soest, S., van Schooneveld, M.J., Riemslag, F.C., de Jong, P.T., Bleeker Wagemakers, E.M. Autosomal recessive retinitis pigmentosa with preserved para-arteriolar retinal pigment epithelium. (1994) *Am. J. Ophthalmol.* **118:** 430–439.

van Soest, S., van den Born, L.I., Gal, A., Farrar, G.J., Bleeker Wagemakers, E.M., Westerveld, A., Humphries, P., Sandkuijl, L.A., Bergen, A.A.B. Assignment of a gene for autosomal recessive retinitis pigmentosa (RP12) to chromosome 1q31-q32.1 in an inbred and genetically heterogeneous disease population. (1994) *Genomics* **22:** 499–504.

Weil, D., Blanchard, S., Kaplan, J., Guilford, P., Gibson, F., Walsh, J., Mburu, P., Varela, A., Levilliers, J., Weston, M.D., *et al*. Defective myosin VIIA gene responsible for Usher syndrome type 1B. (1995) *Nature* **374:** 60–61.

Workman, P.L., Lucarelli, P., Agostino, R., Scarabino, R., Scacchi, R., Carapella, E., Palmarino, R., Bottini, E. Genetic differentiation among Sardinian villages. (1975) *Am. J. Phys. Anthropol.* **43 (2):** 165–176.

Wright, A.F., Mansfield, D.C., Bruford, E.A., Teague, P.W., Thomson, K.L., Riise, R. Genetic studies in autosomal recessive forms of retinitis pigmentosa. In Anderson, R.E., LaVail, M.M., Hollyfield, J.G. (eds) *Degenerative Diseases of the Retina*. pp. 293–302. (1995) Plenum Press, New York.

SECTION VI

◆

FINAL COMMENTS

Chapter 21
Final comments

Elizabeth A. Thompson

◆

In introducing this meeting, Professor Edwards divided us into experts on DNA, experts on lods, and experts on the eye, and pointed to research successes in Oxford in connection with each of the three areas. The meeting has achieved what many attempt: an integration of ideas from all three areas. I believe that all participants, whatever their area of expertise, have learned something about trait complexities of eye disease, the genetic complexities of DNA, and the statistical complexities of lods. The meeting has also been a celebration of the many contributions of Professor Edwards himself, from his linkage work of the 1960s, his work on haplotypes and disequilibria of the 1970s, his work on homologies and associations of the 1980s, and most recently his current contribution in focusing the resolution of complex traits on the search for 'necessary alleles.'

We have moved from foundations, to methods, to tools, to applications, and each of these areas also has aspects concerning DNA, aspects concerning lods, and aspects concerning the eye. The foundations of genetic analysis lie in haplotypes, loci, and alleles. Marker haplotypes constitute the DNA aspect, providing genetic maps, and addressing questions of map resolution and locus order. A model of discrete genetic loci at which marker or phenotypic data are available provides the basis for linkage likelihoods, and hence for lods. The alleles at trait loci determine the trait, such as eye disease, with its many heterogeneities, and lead to questions of the penetrance and specificity of disease-related alleles.

Any method must provide both an approach to inference and an assessment of the uncertainty in any inference. Computation of linkage likelihoods requires a full probability model, with parameters that are to be estimated. The parameters of the model include both parameters of trait determination and

GENETIC MAPPING OF DISEASE GENES
ISBN 0-12-232735-7

linkage parameters such as map locations, and may also incorporate additional population genetic constraints. Methods of linkage detection are necessarily more robust to model assumptions, since they attempt less. Each method entails a choice of the study design or of the subset of available data that will be analyzed. There is no intrinsic difference between analyses of pedigree data, of affected pedigree members, of sib pairs, or of cases and controls from a population. Each design has its associated algorithms or 'methods' for the assessment of the linkage information in the chosen data. Each design has its different advantages in the balance between robustness and power, but all approaches assume the underlying model of a genetic map and trait loci on that map affecting the trait of interest.

Although many methods have been proposed for the linkage analysis of complex traits, there have as yet been few successes. Indeed, for complex traits, linkage estimation must await linkage detection, for only with the additional linkage information will it be possible to resolve the model for trait determination. With the increasing availability of marker data, methods for linkage detection are focusing increasingly on assessment of various forms of association, whether at the population level or among relatives. Methods in the latter category range from analysis of affected sib pairs or more generally affected relatives, to haplotype sharing among affected individuals with presumably more distant unknown genealogical relationships. These methods have had some success for simple traits, but have little power in the absence of Edwards' 'necessary alleles.'

Professor Edwards has exhorted us to consider tools, as foundations and methods require tools for their application. A useful tool for genetic analysis must encompass input, program and output; often the emphasis is on the program, the implementation of an algorithm, with insufficient attention to user interface aspects of input and output. The input must describe the data structure, a pedigree representation or other form, and provide the genetic data on markers and the phenotypic data on traits of interest. A clear and unified input format would go far towards making methods more accessible to practitioners. Equally important is output, with graphical displays of results becoming increasingly widely used.

Finally, we have heard about applications in the area of eye disease, and of some notable successes in resolving traits that at first appear complex in their variability in onset and expression, and, in some cases, difficulty of diagnosis. The successes have been in the areas of genetic and phenotypic heterogeneity, where careful analysis has succeeded in dividing the cases into disjoint genetically homogeneous classes. In many other instances, despite equally careful analysis, the genetic basis of a complex trait remains obscure. It seems such failure of current methods is inevitable for a trait for which (in the words of Professor Edwards) there are common predisposing but 'unnecessary' alleles. The saying of Francis Bacon, quoted by one speaker, is apt: 'Truth emerges more readily from error than from confusion.' The methods developed for the linkage analysis of simple genetic traits have proven highly successful for that purpose. As we begin to understand the failure of these methods to deal with more complex traits, we may develop new methods that are robust, but still have adequate power for linkage detection, at least in the presence of certain kinds of trait complexity. New methods facilitate new applications, but unsuccessful applications also lead to new methods. The integration of theory, algorithms, tools and applications achieved by this meeting will advance both theory and practice.

List of acronyms

A-test	Admixture test
ACEDB	A *Caenorhabditis elegans* Database software
AFM	Association Française Contre Les Myopathies
AGFAP	Antigen genotype frequencies in affected persons
APKD	Adult polycystic kidney disease
APM-method	Affected pedigree member method
ASO	Allele specific oligonucleotide
ASP-method	Affected sib pair method
ASS	Argininosuccinate synthetase
AtDB	*Arabidopsis thaliana* Database
BBS	Bardet–Biedl syndrome
BIDS	Bibliography Information Database System
BLII	British Library Inside Information
BMD	Becker muscular dystrophy
cDNA	Complementary deoxyribonucleic acid
cen	Centromere
CEPH	Centre d'Étude Polymorphisme Humaine
CFTR	Cystic fibrosis transmembrane conductance regulator
cGMP	Cyclic guanosine monophosphate
cM	CentiMorgan
cR	CentiRay
dbEST	Database for ESTs
DF, df	Degrees of freedom
DGGE	Denaturing gradient gel electrophoresis
DKFZ	Deutsches Krebsforschungszentrum
DMD	Duchenne muscular dystrophy
DNA	Deoxyribonucleic acid
E. coli	*Escherichia coli*
EBI	European Biomolecular Institute
EL	Elliptocytosis
EMBASE	Excerpta Medica Database
EMBL	European Molecular Biology Laboratory
ENDRB	Endothelin receptor type B
EPD	Eukaryotic Promotor Database
ERPA	Extended relative-pair analysis
EST	Expressed sequence tag

EU	European Union
EUCIB	European Collaborative Interspecific Backcross Project
FISH	Fluoresent *in-situ* hybridization
FTP	File transfer protocol
GAS	Genetic analysis system
GCG	Genetics Computer Group
GDB	Genome Database
HBD	Homozygous by descent
HGMP-RP	Human Genome Mapping Project – Resource Centre
HHRR	Haplotype-based haplotype relative risk
HLA	Human leucocyte antigen
HRR	Haplotype relative risk
HSCR	Hirschsprung's disease
HSS	Haplotype sharing statistic
IBD, ibd	Identity by descent, identical by descent
IBS, ibs	Identity by state, identical by state
IDDM1	Insulin-dependent diabetes mellitus 1
ISGN	International System for Human Gene Nomenclature
lod	Logarithm of the odds ratio
$\log_{10}$	Logarithm to the base of ten
MASC-method	Marker association segregation chi-square method
MCMC	Monte Carlo Markov Chain
MEN2A, MEN2B	Multiple endocrine neoplasia type 2a and 2b
MITF	Microphthalmia-associated transcription factor
MLS-test	Maximum likelihood score test
Mlod	Maximum lod over segregation and recombination
MRC	Medical Research Council
MTC	Medullary thyroid carcinoma
MycDB	*Mycobacterium* Database
NPL	Nonparametric linkage
NRL	Neural retina leucine zipper
OMIM	On-line mendelian inheritance in man
p.d.f.	Probability distribution function
PAC	P1 artificial chromosome
PCR	Polymerase chain reaction
PIR	Protein information resource
PPRPE	Para-arteriolar preservation of the retinal pigment epithelium
PS-test	Predivided sample test
pter	Short arm telomere of a chromosome
RDS	Retinal degeneration: slow
RET	RET Oncogene

RFLP	Restriction fragment length polymorphism
Rh	Rhesus blood group
RNA	Ribonucleic acid
RP	Retinitis pigmentosa
RR	Relative risk
Sim	Simulation-based statistics
SRS	Sequence retrieval system
SSCP	Single strand conformation polymorphism
TDT	Transmission/disequilibrium test
TFD	Transcription factor database
TS two loci	Tuberous sclerosis
URL	Uniform relocation link
USDA	United States Department of Agriculture
USH	Usher's syndrome (several forms)
VAPSE	Variant affecting protein structure or expression
WS	Waardenburg's syndrome (several forms)
WWW	World wide web
YAC	Yeast artificial chromosome

Index

A

Abetalipoproteinaemia, 257
Additive method, 10, 12
Admixture test, 99–105
Adrenal hyperplasia, 46
Adult polycystic kidney disease (APKD), 97
Affected pedigree member (APM) method,
    108–9, 144, 148–9, 216, 276
    simAPM, 109, 189, 190–1, 193–203
Affected sib pair (ASP) method, 26–7, 106–8,
    147–56, 276
    compared to other nonparametric methods,
        109, 189, 190, 203
    MASC-method, 182
    recessive disease, 32, 33, 36–7, 38, 39–42,
        46, 47–55
    XYZ test, 175
Aganglionic megacolon (Hirschsprung's
    disease), 239–44
Age, of onset, heterogeneity analysis, 104–5
Albinism, 4
Allele-specific oligonucleotides (ASOs), 250
Allelic association *see* Association studies
Allelic heterogeneity, 197
Alzheimer's disease, 109, 142, 148
ANALYZE, 216
Aniridia, 36
Ankylosing spondylitis, 161
Antigen genotype frequencies in affected persons
    (AGFAP), method, 175, 182, 183–4
A *Arabidopsis thaliana* Database (AtDB), 212
Argininosuccinate synthetase (ASS), 78, 84
Association studies, 4, 19–20, 159–70, 275
    and linkage information in multifactorial
        disease, 179–86
    compared linkage analysis, 172–7
    haplotype sharing, 116, 119, 120, 130
    Hirschsprung's disease, 243, 244
    identity by descent, 137, 139–40, 144
    retinitis pigmentosa, 265
    see also Recessive disease

Assumed mode, lod, 152
Asthma, 132, 147, 265
A-test (Admixture test), 99–105
Attributable risk (AR), 167–9
Autosomal inheritance, mutation-selection-
    equilibria, 66–71
    dominant inheritance, 68–9
    dominant inheritance with incomplete
        penetrance, 70
    recessive inheritance, 70–1
    see also Retinitis pigmentosa

B

Band, map integration, 19
Bardet-Biedl syndrome (BBS), 255, 257, 261–4
Batten's disease, 34, 35
Bayes Theorem, 89–90
Becker muscular dystrophy (BMD), 64–5
Benign recurrent intrahepatic cholestasis, 116
β Model, 18
Between-family information, 32, 44
Bibliographic Services, 211, 213–4
Bibliography Information Database System
    (BIDS), 213
Bimodal, retinal dystrophy, 251
Bipolar affective disorder, 100, 148
Bipolar I disorder, 108
Bonferoni correction, 170
British Library Inside Information (BLII),
    213–4
B-test, 99

C

Caenorhabditis elegans, 6
A *Caenorhabditis elegans* Database (ACEDB),
    210, 212, 216, 221–34
Cancer, 23
Candidate gene, 43, 160, 173–7, 179–86
    Hirschsprung's disease, 240
    retinitis pigmentosa, 256–60, 267
Cartilage-hair hypoplasia, 116

283